John Berkenhout

Clavis Anglíca Linguæ Botanicæ

A Botanical Lexicon

John Berkenhout

Clavis Anglíca Linguæ Botanicæ
A Botanical Lexicon

ISBN/EAN: 9783337225865

Printed in Europe, USA, Canada, Australia, Japan

More available books at **www.hansebooks.com**

LINGUÆ BOTANICÆ;

OR, A

BOTANICAL LEXICON;

IN WHICH

The Terms of Botany, particularly thofe occurring in the Works of Linnæus, and other modern Writers,

ARE

APPLIED, DERIVED, EXPLAINED, CONTRASTED, and EXEMPLIFIED.

1 Kings iv. 33.

וַיְדַבֵּר עַל־הָעֵצִים מִן־הָאֶרֶז אֲשֶׁר בַּלְּבָנוֹן
וְעַד הָאֵזוֹב אֲשֶׁר יֹצֵא בַּקִּיר

LONDON:

Printed for the Author.
Sold by T. Becket, and A. de Hondt, in the *Strand*; and Meff. Hawes, Clarke, and Collins, in *Pater-nofter-row*.

MDCCLXIV.

T O

JOHN HOPE, M. D.

Fellow of the ROYAL COLLEGE
of PHYSICIANS, and PROFESSOR
of BOTANY and MEDICINE in
the Univerfity of Edinburgh.

S I R,

IF I have made any progrefs in the
ufeful and amufing ftudy of Botany, it muft be chiefly attributed
to the inftruction which I imbibed
from your lectures; to whom therefore, could I, with equal propriety,
infcribe this volume, as to him, who,
in reality, is the caufe of its exiftence?

COMMONLY the defign of an author, in his dedication, is to exhibit to the world a flattering portrait of his patron: a cuftom proceeding probably from a perfuafion, that mankind in general are delighted with praife, and *not offended* by adulation; or from a fuppofition, that the virtues, real or imaginary, of the perfonage, under whofe aufpices the author appears, will reflect fome luftre on himfelf. Whatfoever may be his motive, he is certainly miftaken in the effect; by afcribing fuch fuperlative excellence to humanity, he exceeds the limits of human nature, and, inftead of a juft refemblance, prefents us with a *perfect* monfter. The natural confequence, however, of this prevalent cuftom is, that mankind are taught to

regard

regard the praife of a dedication as mere words of courfe, which cannot therefore in any degree exalt their ideas either of the author or his patron.

THESE confiderations are fufficient to overpower my inclination to expatiate, as with truth I might, on your improvement of the fcience of Botany in this univerfity, and on what may be ftill expected from your knowledge, affiduity, and inclination : with regard particularly to the laft, the gold medal, with which you are pleafed annually to reward the ftudent who prefents the beft *hortus ficcus*, is a better teftimony than any thing I could fay upon the fubject.

SINCE

SINCE the general admiffion of the Linnæan fyftem, an explanation of botanical terms, in the form of a dictionary, feems to have been univerfally defired: it is therefore wonderful that none of our noted Botanifts fhould hitherto have obliged us in this particular. For want of fuch a work, a Profeffor of Botany is under a neceffity of devoting a confiderable part of his courfe to the mere explanation of words; a part which is certainly the leaft agreeable to himfelf, and leaft entertaining to his pupils: if this Lexicon fhould be found, in any degree, to fuperfede that neceffity, I fhall think myfelf fufficiently rewarded for my labour.

I wish it had been in my power to have rendered it lefs imperfect, and confequently lefs unworthy your acceptance; but my prefent indifpenfable application to a ftudy of which Botany is but a branch, obliged me to finifh the work in lefs time than I could have wifhed; neverthelefs, I flatter myfelf that you will not find many material omiffions, or capital miftakes.

Your permiffion to prefix your name to this epiftle, is a proof that you think the book may be of fome utility, which cannot fail to fecure me a favourable reception from the public; for this permiffion therefore, I thank you moft fincerely, and in

the

the flattering perfuafion that you will continue to honour me with your friendfhip, beg leave to fubfcribe myfelf,.

Your moft obedient;

and humble fervant;

Edinburgh, }
Jan. 1. 1764. }

JOHN BERKENHOUT.

PREFACE.

WHEN I began the study of Botany, I could not help lamenting the want of an ample explanation of botanical terms, arranged and digested in some such manner as in the volume which I herewith presume to offer to the public, as it was easy to conceive that such an assistant would greatly facilitate the study in which I had engaged. I sought in vain among the numerous tribe of our technical dictionaries for an explanation of a language, which, in a great measure, owes its existence to Linnæus, the father of modern Botany. That distinguished naturalist has, indeed, favoured us with a Latin explanation of many of his terms in his *Philosophia Botanica*; but, since the publication of that work, he has added a very considerable number of words, used in a

b   sense

fenfe peculiar to himfelf, which remain hitherto unexplained. Part of the *Philofophia Botanica* has been tranflated into Englifh, under the title of an *Introduction to Botany*; but that tranflation is lefs ufeful than it might have been, if the tranflator had not given himfelf the trouble to *anglicife* (if I may be allowed the expreffion) the terms themfelves; a very ufelefs attempt, as there is hardly a fingle botanical book, of any repute, in the Englifh language; Latin is the eftablifhed language of Botany in all nations.

THERE is perhaps no circumftance which would be of greater advantage to the fcience of Botany in particular, than that of fixing an abfolute fignification to all its terms : this is hardly to be expected in a firft attempt; but there is a poffibility that fuch an attempt may become the foundation of a more perfect fuperftructure.

ture. With this view I take the liberty to requeft of the more experienced botanifts, that where-ever they find me to have erred in the explanation of a term, they will pleafe to communicate their correction in a line, directed for me, to the care of the publifhers; and, how difpleafed foever mankind in general may be when fhewn their errours, in this particular inftance I affure them, that I fhall very fincerely acknowledge the obligation. There are fome few words, the precife meaning of which, I own, I do not comprehend; as for example, *Acutum* & *Obtufum*, applied to *Perianthium*. Of thefe, efpecially the firft, the reader may find repeated examples in the clafs *Pentandria*, in the *Genera Planta-rum*: concerning thefe in particular, I fhould be glad to be better informed. If they do not refer to the fhape of the *Calyx* before the expanfion of the flower, I fee

no

no meaning in them at all ; and yet there are fome objections to this explication.

I cannot, in juftice, neglect to acknow-ledge my obligation to my worthy friend Mr ARTHUR LEE *, for his kind affift-ance ; a gentleman who will be a fingular credit to this univerfity, and a bleffing to to that country in which he fhall hereafter practife the healing art.

THE reader will eafily do me the juftice to believe, that vanity could have no in-fluence in perfuading me to the publica-tion of a work, the compofition of which required neither genius nor learning.

---

* A native of Virginia ; he obtained Dr HOPE's prize-medal, for the beft *hortus ficcus* in the year 1763.

CLAVIS

CLAVIS ANGLICA

LINGUÆ BOTANICÆ.

ABBREVIATUM *Periantbium,* when the *Perianthium* is shorter than the tube of the *Corolla,* opposed to *Longum*; exemplified in the *Pulmonaria maritima.*

ABORTIENS *Flos,* a term used by former botanists; *Sterilis* of Tournefort; *Masculus* of Linnæus; *Paleaceus* of Ray. See *Masculus.* By *Abortiens, Sterilis,* &c. former botanists meant such flowers as produced no fruit. Linnæus, finding this to be generally owing to their being male flowers, changed the term to *masculus*; but in the class *Dioecia* it will often be the case with female flowers, if they have no male near. See *Phil. Transf. vol.* 47. *p.* 169.

ABRUPTUM *Folium pinnatum* [ex *abrum-*

A   *por,*

por, to be broken] *terminatum neque cirrho, neque foliolo*; ending abruptly without either tendril or leaf.

ACAULIS *Herba* [à priv. & *caulis*] without stem, oppofed to *Caulefcens*.

ACEROSUM *Folium* [ex *acus*, chaff] *eft lineare perfiftens*, linear and perfifting; furrounded at the bafe by chaffy *fquamæ*, as in the *Pinus, Abies, Juniperus, Taxus*.

ACICULARIS [ab *acicula*, a pin, or fmall needle] fmall and fharp pointed. The trivial name of a fpecies of the *Scirpus*.

ACINACIFORME [*acinaces*, a Perfian fcymitar] *eft compreffum carnofum, altero margine convexo angufto, altero rectiore craffiore.* This term implies fubftance, one edge of the leaf being convex and fharp, and the other rather ftraighter and thicker, as in the *Mefembryanthemum acinaciforme.*

ACINI, the fmall berries which compofe a mulberry, the berry of the bramble, &c.

ACOTYLEDONES [à priv. & *Cotyl.*] A term of Placentation, applied to thofe plants whofe whose seeds have no Cotyledons, as in the *Musci.* See *Cotyledon.*

ACULEI [ab 'Aκις, *cuspis*, a point] prickles; a species of *Arma* on the surface of some plants given them for their defence against certain animals, as in the *Volkameria, Pisonia, Hugonia, Cæsalpina, Mimosa, Parkinsonia,* &c. *Aculei.* are either *recti, incurvi,* or *recurvi,* and are fixed only in the rind, so as to be separable from the plant without tearing its substance.

ACULEATUS *Caulis, Folium* [ab *aculeus,* a sting] beset with stiff, sharp prickles; between *hispidus* and *spinosus: cum acumina pungentia rigida occupant discum :* furnished with *Aculei,* which see.

ACUMINATUM *Folium* [ab *acuo,* to sharpen] terminating in a long tapering point, *quod terminatur in apice subulato.*
Acuminatus Calyx, as in the *Itea.*

ACUTUM *Folium* [*acuo,* to whet]. *quod terminatur angulo acuto,* says Linnæus, *i. e.* terminating in an acute angle, different from *acuminatum,* in not running out into a subulated point.
Acutum Perianthium, as in the *Primula, Androsace, Dactylis, Conocarpus, Campanula,* &c.

AD-

ADNATUM *Folium* [*ad*, & *nafcor*, to be born, to grow] growing clofe to the ftem. I conceive no difference between this term and *Adpreffum*, unlefs it implies adhefion.

Adnatæ Stipulæ, growing clofe to the plant, oppofed to *Solutæ* ; exemplified in the *Rofa*, *Rubus*, *Potentilla*, *Comarum*, *Melianthus*.

Adnatus Stylus, adhering to the *Corolla*, as in the *Canna*.

ADPRESSA *Folia* [*ad*, to, & *preffus*, preffed] the difk of the leaves approaching the ftem fo as almoft to touch it ; *dum difcus folii approximatur cauli*, fays Linnæus, in his explanation ; but the word itfelf properly implies being preffed clofe to the ftem.

ADSCENDENS *Caulis* ; afcending, *i. e.* growing firft in a horizontal direction, and then gradually curving upwards.

ADVERSUM *Folium* ; *quod latus meridiei obvertit*, turned towards the fouth, as are thofe of the *Amomum*.

ADULTERINUS, baftard, as *Acorus adulterinus*, baftard *Acorus*, the *Iris pfeudacorus* fo called by Bauhinus : fynon. with *Pfeudo*.

ÆQUA-

ÆQUALIS *Polygamia*, equal, The firſt order in the claſs *Syngenefia* of Linnæus, con- fiſting of thoſe plants in the compoſition of whoſe flowers all the *floſculi* are hermaphro- dite.

Æqualis Corolla, equal; *i. e.* where the parts are equal as to figure, magnitude, and propor- tion, as in the *Primula, Limoſella.*

Æqualis Calyx, as in the *Utricularia.*

AGGREGATUS *Flos*, [*aggrego*, to aſſem- ble.] Flowers are called *Aggregate*, when, by means of ſome part of the fructification, many *Floſculi* are ſo united that none of them could be taken away without deſtroying the uniformi- ty of the whole. The part which in aggregate flowers is common to the whole, is either the *Receptaculum* or *Calyx*. Aggregate flowers are principally divided into ſeven kinds, viz. *Um- bellatus, Cymoſus, Compoſitus, Aggregatus* proper- ly ſo called, *Amentaceus, Glumoſus, Spadiceus. Aggregatus Flos*, properly ſo called, is that which has a dilated *Receptaculum*, with *Floſculi* ſtanding on *Pedunculi*. For *Aggregata Radix*, ſee *Granulata*.

Aggregatæ ſtellares Pubeſcentiæ, a ſpecies of ſetaceous pubeſcence, on the ſurface of ſome plants, in which the larger *ſtellæ* produce ſmal- ler

ler ones on the *apex* of each *seta*, as in the *A-lyssum*, *Helicteres*.

Aggregatæ, an order of plants in the *Fragmenta methodi naturalis* of Linnæus, containing these *genera*, viz. *Statice*, *Protea*, *Leucadendros*, *Hebenstretia*, *Brunia*, *Cephalanthus*, &c.

ALA, a wing ; the two lateral *petala* of a papilionaceous *corolla* ; also a membrane fixed to some species of seeds by which they fly and disperse. Used by former botanists to express the angle formed by the stem and branch or leaf. See *Axillare*.

ALARIS *Pedunculus* [*Ala*] See *Axillaris*.

ALATUS *Petiolus* [*ala*, a wing] winged ; not linear, but spreading to each side with little membranous wings, as in *Aurantium*, and *Hedysarum*.

ALBA *Lactescentia*, white, as in the *Euphorbia*, *Papaver*, *Asclipias*, *Apocynum*, *Cynanchum*, *Campanula*, & *Semiflosculosi* of Tournefort.

ALBURNUM [*albus*, white] The white substance which lies between the inner bark and the wood, in trees ; called by some *Adeps arborum :* it is closely connected with the *Liber*,

and

and to be feparated from it with difficulty.

ALGÆ, one of the feven families, or tribes, in the vegetable kingdom, defined by Linnæus to be fuch as have their root, leaves, and *caudex* all in one, comprehending fea-weeds, and fome other aquatic plants. In Tournefort they conftitute the fecond *genus* of the 2d fection of clafs xvii. and are divided into nine fpecies. In the *Syftema Naturæ* of Linnæus they conftitute the third order in the clafs *Cryptogamia*, are divided into *terreftres*, and *aquaticæ*, the firft comprehending eight *genera*, and the latter fix.

Algæ, an order of plants in the *Fragmenta methodi naturalis* of Linnæus.

ALTERNATIVÆ *petiolares Gemmæ* (for *alternæ*) alternate; as in the *Salix, Spiræa, Genifta, Solanum, Ilix, Juglans,* &c.

Alternativæ ftipulaceæ Gemmæ, as in the *Populus, Tilia, Ulmus, Quercus*, &c.

Alternativæ ftipulaceo-petiolares Gemmæ, as in the *Sorbus, Crataegus, Prunus,* &c.

ALTERNI *Rami, Folia*; alternate branches, leaves, &c. oppofed to *oppofiti*; growing not oppofite to each other, *cum unum poft alterum tanquam per gradus exit.*

AMEN-

AMENTACEÆ [*Amentum*] An order of plants in the *Fragmenta methodi naturalis* of Linnæus, containing the following *genera*, viz. *Piſtacia*, *Myrica*, *Alnus*, *Betula*, *Salix*, *Populus*, *Platanus*, *Carpinus*, *Corylus*, *Juglans*, *Quercus*, *Fagus*.

Amentaceus Flos, an aggregate flower having a *receptaculum filiforme* with amentaceous ſcales.

AMENTUM [ab ἄμμεϛ, *vinculum*, a bond or thong] the *Calyx* ſo called, when, proceeding from a common receptacle, it is alternately mixed with the flowers, ſomewhat like the chaff in an ear of corn. Linnæus defines it by the compound word *paleaceo-gemmaceo*. It is termed by former botaniſts a *Julus*, and in Engliſh a *Catkin*. It occurs frequently in the claſs *Monoecia*. It is the *Nucamentum*, and *Catulus* of ſome writers.

AMPLEXICAULE *Folium* [*amplector*, to embrace, and *caulis*, a ſtem] *ſi baſis folii undique ambiat latera caulis tranſverſim*, the *baſis* of the leaf entirely ſurrounding the ſtem tranſverſely. Diſtinguiſhed from *Vaginans* by the word *tranſverſim* : exemplified in the *Potamogeton perfoliatum*, *Verbaſcum blattaria*, *Hyoſcyamus niger*.

ANCEPS *Caulis*, *two-edged*; forming two oppofite acute angles, of which the *Sifyrinchium* is
an example. The *Anceps* may have many more
angles, but then they will be all obtufe, except
the two oppofite ones which conftitute the *anceps*. When applied to a leaf, it implies fubftance, and fignifies its having two oppofite
longitudinal angles with a convex difk.

ANDROGYNA *Planta* [ex ἀνὴϱ, *vir*, a
man, & γυνὴ, *mulier*, a woman] fuch plants as
bear both male and female flowers on the fame
root, as in the clafs *Monoecia*.

ANGULATUS 3—10 *Caulis:* angulated,
oppofed to *teres*, *femiteres*, *compreffus*, &c.

ANGUSTIFOLIA [*anguftus*, narrow, & *folium*, a leaf]. Narrow leaved.

ANGYOSPERMIA [Αγγσ, *vas*, a veffel].
The fecond order in the clafs *Didynamia* of
Linnæus: it confifts of thofe plants, of that
clafs, whofe feeds are inclofed in a *Pericarpium*.
In this order the ftigma is generally obtufe.
Thefe are the *Perfonati* of Tournefort.

ANNUA *Radix*, [ab *annus*, a year] an an-
nual root; that which lives but one year.

B . ANOMALÆ

ANOMALÆ *Gemmæ* [ex à *priv. & ομαλὸς, æqualis*] irregular, as in the *Abies*, *Pinus*, *Taxus*, &c. oppofed to *Oppofitivæ*, and *Alternativæ*.

ANTHERA [from Aνθ☉, *flos*, a flower] that part of the *Stamen* which is fixed on the top of the *Filamentum*, within the *Corolla*; it contains the *Pollen*, or fine duft, which, when mature, it emits for the impregnation of the plant, according to Linnæus. *Antheræ* are either *diftinctæ*, *connatæ*; *loculi*, or *apertura*. The *Apex* of Ray, Tournef. & Rivin. *Capfula ftaminis*, of Malpigh.

APERTURA, an aperture; the minute opening in fome fpecies of *Antheræ*.

APETALUS *Flos* [*a*, priv. & *petalum*]. Having no corolla. *Stamineus*, Ray; *Incompletus*, Vaillant; *Imperfectus*, & *Capillaceus*, other Botanifts: exemplified in the *Lepidium ruderale*.

APEX *Folii* [dict. ab *apiendo*, i. e. *ligando*] the top or fummit; the upper extremity of the leaf oppofite to the *bafis*. A leaf, refpecting its apex, may be *truncatum*, *præmorfum*, *retufum*, *emarginatum*, *obtufum*, *acutum*, *acuminatum*, or *cirrhofum*.

APHYLLUS

APHYLLUS *Caulis*; [from *a*, and Φυλλον, *folium*, a leaf] deftitute of leaves.

APOPHYSIS [ab απο, & φυο, *nafcor*, to grow from] an excrefcence from the *Receptaculum* of the *Mufci*: it is *marginata* in the *Sphagnum*.

APPENDICULATUS *Petiolus* [*appendicula*, dim. ab *appendix*, a little appendage] hanging at the extremity of the ftem.

APPROXIMATA *Folia*; leaves growing near each other, oppofed to *Remota*.

ARBOR, a tree. Trees are by Linnæus claffed in the feventh family of the vegetable kingdom, and are diftinguifhed from fhrubs in that their ftems come up with buds on them: but this diftinction holds not univerfally, there being rarely any buds on the large trees in India. According to Ludwig, *Arbor eft planta quæ truncum fimplicem et lignofum habet.*

ARBOREUS *Caulis* [*Arbor*, a tree] fimple, ligneous, and continuing; oppofed to *fruticofus*, *fuffruticofus*, & *herbaceus*.

ARBUSTIVA [*Arbuftum*, a copfe of fhrubs, or trees; an orchard, a vineyard]. An order

of

of plants in the *Fragmenta methodi naturalis* of Linnæus, in which are thefe *genera*, viz. *Philadelphus, Eugenia, Pfidium, Myrtus, Caryophyllus.*

ARCUATUM *legumen* [ab *arcus*, the curvature of an arch or of a bow-ftick] curved, as in the *Ornithopus perpufillum.*

ARECTUM *Folium.* See *Erectum.*

ARILLUS, the proper exterior coat of a feed which falls off fpontaneoufly : it is exemplified in *Coffea, Jafminum, Cynogloffum, Cucumis, Dictamnus, Diofma, Celaftrus, Euonymus.* The *Arillus* is either *cartilagineus,* or *fucculentus.*

ARISTA [ab *areo*, to be dry or parched] the beard of corn, or grafs iffuing from a *gluma.*

ARISTATA *Gluma* [*Arifta*] having an *arifta*, oppofed to *mutica.*

ARMA, arms, weapons ; one of the feven kinds of *Fulcra* of plants, according to Linnæus, intended by nature to fecure them againft external injury : its fpecies are, *Aculei, Furcæ, Spinæ, Stimuli.*

AR-

ARTICULATUS *Caulis, Culmus*; having knots or joints.

Articulata Radix, a jointed root, as in *Lathræa, Oxalis, Martinia, Dentaria.*

Articulata folia, cum folium unum ex alterius apice excrescit, attached to the ends of each other, resembling the links of a chain.

Articulate folium pinnatum, when the *foliola* are attached to the extremities of each other, proceeding from one common *petiolus.*

ARTICULUS *Culmi* [ab *artus*, a joint or limb] the straight part of the *Culmus* between two *Geniculi.*

ASCYROYDEÆ [ab Ασχυρον, Pliny's name for the *Hypericum*]. The thirteenth natural class in Scopoli's *Flora Carniolica.*

ASPERIFOLIÆ [*asper*, rough, & *folium*, a leaf]. An order of plants in the *Fragmenta methodi naturalis* of Linnæus, in which are these genera, viz. *Tournefortia, Cerinthe, Symphytum, Pulmonaria, Anchusa, Lithospermum, Myosotis, Heliotropium, Cynoglossum, Asperugo, Lycopsis, Echium, Borrago:* magis minusve oleraceæ, mucilaginosæ, & glutinosæ sunt. *Lin.* In the present system, these are among the *Pentandria monogynia.*

ASSURGENTIA *Folia* [*aſſurgo*, to riſe up] *arcuatim erecta* ; firſt declining, but growing e-rect towards the *apex*.

ATTENUATUS *Pedunculus* [*attenuor*, to be waſted, worn] when gradually ſmaller to-wards the flower, oppoſed to *Incraſſatus*.

AUCTUS *Calyx* [ab *augeor*, to be increaſed] when a ſeries of ſhorter and different *ſquammæ* or *ſquammulæ* ſurround the exterior *baſis* of the *Calyx*, as in *Coreſpſis*, *Bidens*, *Crepis*, *Dianthus*. Linnæus defines the *Calyx auctus* in the claſs *Syn-geneſia*, thus ; *dum unica ſeries laciniarum æqualis longior cingit floſculos, &. alia minima cingit baſin tantum ſimplicis interioris & majoris calycis.*

AVENIA *Folia* [*a*, & *vena*, a vein] leaves which have no viſible veins. Vid. *Venoſa*.

AURICULATUM *foliolum* [ab *auricula*, a little ear, dim. ab *auris*, the ear] twiſted into the form of a little ear, exemplified in the *Jun-germannia ciliaris.*

AXILLARIA *Folia* [*Axilla*, the arm-pit] growing out of the angles formed by the branches and the ſtem. The ſame as *Subalaria*.
Axillaris Pedunculus, proceeding from the *axil-*
la

la formed by leaves or branches with the ftem, as in the *Meliffa calamintha*, *Nepeta*, and many other flowers.

B.

BACCA, a berry; a full, pulpy *Pericarpium*, without *Valvulæ*, in which the feeds are naked, having no other covering or cell, as in the goofberry, *&c.*

BARBA, a beard; a fpecies of pubefcence covering the furface of plants; it does not appear in the *Phil. Botanica*, and therefore remains unexplained. In the *Delineatio Plantæ* it is ranged thus, *Pili, Lana, Barba, Fomentum.* It feems from its application in the *Spec. Pl.* to fignify a tuft of hair, *&c. &c.*

BARBATUM *Folium* [*barba*, a beard]. If Linnæus intends that this term, applied to the furface of a leaf, fhould have a precife meaning diftinct from *pilofum, hirfutum, villofum*, it muft certainly allude to the beard of a goat, *i. e.* the hairs ending in a point.

Barbatus Flos, inftanced in the *Dianthus barbatus*, Sweet William.

Barbatus Corolla, in the *Gentiana camp.*

BICORNES [*bis*, & *cornu*, a horn]. An order of plants in the *Fragmenta methodi naturalis* of Linnæus, in which are the following *genera*, viz. *Ledum, Azalea, Andromeda, Clethra, Erica, Myrfine, Memecylum, Santalum; Vaccinium, Arbutus,* &c. Adftringunt, fed baccæ acidæ efculentæ funt. *Lin.* Thefe are fo termed from the *Antheræ* having in appearance two horns.

BIENNIS *Radix* [ex *bis*, twice, & *annus*; a year] a root which continues to vegetate two years.

BIFARIA *Folia* [*bis*, & *fari*, to fpeak] pointing two ways.

BIFERÆ *Plantæ* [*bis*, & *fero*, to bear] flowering twice a-year, fpring and autumn, common between the tropics.

BIFIDUM *Folium* [ex *bis*, twice, & *Fiffum*, cloven] twice divided, its finufes linear and margins ftraight. See *Fiffum.*

BIFLORUS *Pedunculus* [*bis*, & *flos*, a flower] bearing two flowers; producing two fructifications upon each *Pedunculus.*

BIGEMINUM *Folium compofitum* [*bis*, twice, &

& *geminus*, double] a forked *petiolus* with two *foliola* on the *apex* of each diviſion, *cum petiolus dichotomus apicibus adneɛtit foliola quatuor.*

BIJUGUM *Folium* [*bis*, & *jugo*, to yoke] a pinnate leaf conſiſting of two pair of *foliola*.

BILABIATUS *Corolla* [*bis*, & *labium*, a lip]. A *Corolla* with two lips, as the *Pinguicula*, and moſt of the *Dydinammia*.

BILOBUM *Folium* [ex *bis*, twice, & Λοϐός, *the tip of the ear*] conſiſting of two lobes. See *Lobatum.*

BINATA *Folia* [à *binus*, two and two] indicating the number of *foliola* in a *folium digitatum* ; conſiſting only of two *foliola*.
Bini Pedunculi, growing in pairs, as in *Capraria*, & *Oldenlandia Zeylanica.*

BIPARTITUM *Folium* [*bis*, & *partitus*, divided] conſiſting of two diviſions *uſque ad baſin*, down to the baſe.

BIPINNATUM *Folium compoſitum* [*bis*, & *pinnatum*, winged] doubly winged ; *cum petiolus lateribus adfigit foliola pinnata*, i. e. when a *petiolus* is pinnated by lateral *petioli*, which are

C   themſelves

themselves pinnated by *foliola*, as in the *Atha-manta libanotis, Anemone pulsatilla.*

BITERNATUM *Folium compositum* [*bis*, twice, & *ternus*, threefold] a *petiolus* with three divi-fions, and three *foliola* upon each ; *duplicato ter-natum, cum petiolus adfigit tria foliola ternata,* as in the *Epimedium,* & *Ligusticum scoticum.*

BIVALVE *Pericarpium* [*bis*, & *valvæ*, doors or valves] confifting of two valves, as the *Sili-qua* & *Legumen,* which fee.

BLATTARIÆ [à *Blatta*, a moth, or little worm] the title of Scopoli's twelfth natural clafs, in his *Flora Carniolica* ; it is taken from the *Blattaria* which was Tournefort's generic name for the *Verbafcum* of Linnæus.

BRACHIATUS *Caulis* [*Brachium,* an arm] having branches, in pairs, oppofite to each o-ther, each pair ftanding at right angles with thofe above and below.

BRACHIUM, the Arm. The tenth degree in the Linnæan fcale for meafuring plants : from the *Axilla* to the extremity of the middle fin-ger ; or twenty-four Parifian inches.

BRACTEA,

BRACTEA, a thin leaf or plate of any metal; *folium florale*, ranged by Linnæus among the *Fulcra* of plants. Thefe floral leaves differ in fhape and colour from the other *folia* of the plant, are generally fituated on the *pedunculus*, and often fo near the *corolla* as to be eafily miftaken for the *calyx*, than which however the *Bractea* are generally more permanent. Examples of floral leaves are feen in the *Tilia, Fumaria bulbofa, Lavendula, Horminum.* *Bractea* are either *coloratæ, caducæ, deciduæ, perfiftentes; una, duæ, plures; coma; foliorum cetera addenda.*

BRACTEATUS *Pedunculus*, [*bractea*, a floral leaf] having *bractea* growing on it.

BULBIFERUS *Caulis* [à *Bulbus*, a round root] bearing bulbs : thefe are generally on the defcending *caudex* ; but when on the *caudex afcendens*, if they touch the ground, they immediately put forth *fibrillæ*, and become real roots, as in the *Ranunculus ficaria.*

BULBOSA *Radix* [à *Bulbus*, a fpecies of onion] enlarging in a globular form at the bottom of the afcending caudex, and fhooting forth *radiculæ* from its bafis. A bulbous root is either *fquammofa, tunicata, duplicata, folida, or articulata.*

C 2   BULBUS,

BULBUS, a species of *Hybernaculum* on the *caudex descendens*.

BULLATUM *Folium* [*bulla*, a bubble] when the substance of the leaf rises high above the veins so as to appear like little blisters; *rugosum* in a greater degree.

C.

CADUCUM *Folium* [à *cado*, to fall] a term signifying the shortest time of duration; falling off at the first opening of the flower.

Caducus Calyx, as in the *Papaver* & *Epimedium*.

CALAMARIÆ [*Calamus*, a reed]. An order of plants in the *Fragmenta methodi naturalis* of Linnæus, in which he has these genera, viz. *Bobartia, Scirpus, Cyperus, Eriophorum, Carex, Schoenus, Flagellaria, Juncus.*

CALCARIATUM *Nectarium* [*Calcar*, a spur] In shape resembling a cock's spur, as in the Lark's spur, the *Antirrhinum, Valeriana, Pinguicula, Utricularia. Calcar est nectarium ex corolla pone in conum extensa*: When applied to *Corolla*, it relates to the *Nectarium*.

CALI-

CALICULATUS *Calyx* [*Calicula*, dim. à *Calyx*] having its *basis* inclofed within a fmall exterior *calyx* ; fynon. with *Auctus* ; *Completus* of Vaillant; exemplified in the *Leontice leontope-taloid*, *Prenanthes*.

CALYCANTHEMI [*Calyx*]. An order of plants in the *Fragmenta methodi naturalis* of Linnæus, in which are thefe genera, viz. *Epilobium, Oenothera, Juffiæa, Ludwigia, Oldenlandia, Ifnarda,* &c.

CALYCIFIBRÆ [à *Calyx*, & *fibra*, a fibre]. A natural clafs in Scopoli's *Flora Carniolica*.

CALYCIFLORÆ [*Calyx* & *flos*]. The eleventh clafs in Royen's fyftem : it is in fact the *Floribundæ* of Linnæus's *Methodus Calycina*. The fecond, third, and fourth order are taken from the *Icofandria* in the fexual fyftem.

CALYPTRA [from Καλύπτω, *tego*, to cover] a veil ; the *Calyx* of moffes, covering the *Antheræ* like a hood : it may be *recta*, or *obliqua*. Ufed by former botanifts to exprefs that which Linnæus calls the *Arillus*.

CALYX [ex Καλύπτω, *tego*, to cover] the firft of the feven parts of fructification, according
to

to Linnæus, and by him defined to be the outer bark of the plant prefent in the fructification. In general, it is that green cup which inclofes and fupports the bottom of the *Corolla*, and is otherwife called *Perianthium, Involucrum, Amentum, Spatha, Gluma, Calyptra,* or *Valva,* as it happens to be differently circumftanced. In affimulating the vegetable with the animal kingdom, Linnæus terms the Calyx *florum thalamus*. It is generally fingle, in fome plants double and in others entirely wanting. It is commonly divided into the fame number of fegments with the *Corolla*. The *Calyx* commonly withers when the fruit is ripe, if not before; which circumftance infallibly diftinguifhes the *Calyx* from *Bractea*, in dubious cafes. It is generally lefs, in point of height, but more fubftantial, than the *Corolla*.

CAMPANACEI [*Campana*, a bell]. An order of plants in the *Fragmenta methodi naturalis* of Linnæus, in which are the following *genera*, viz. *Convolvulus, Ipomoea, Polemonium, Campanula, Roella, Viola,* &c.

CAMPANULATUS *Corolla* [à *Campanula*, a little bell] fhaped like a bell, having no tubular bafis, *ventricofus abfque tubo*, as in the *Campanula, Convolvulus, Atropa,* and feveral fpecies of the *Gentiana*.

CANA

CANALICULATUM *Folium* [*Canalicula*, dim. à *canalis*, a channel] having a deep channel running from the base to the apex ; *ex sul-co profundo, secundum totam longitudinem, excavatum in dimidiatum fere cylindrum.*

CANDELARES [*Candela*, a candle]. An order of plants in the *Fragmenta methodi natura-lis* of Linnæus, containing these genera, viz. *Rhizophora, Mimusops, Nyssa.*

CAPILLACEUM *folium*, Capillary, the same as *capillare*, exemplified in the *Jungerman-nia rupestris, Ranunculus aquatilis.*
Capillacea radix, a species of the fibrous root, exemplified in the *Gramina.* Ludwig.

CAPILLARIS *Pappus* [*capillus*, hair] simple and filiform, as in the *Hieracium, Son-chus*, &c.
Capillares Glandulæ, resembling hairs, as in the *Ribes, Antirrhinum quadrifolium, Scrophula-ria, Cerastium, Silene.*

CAPILLUS [qu. *capitis pilus*] hair. The first degree in the Linnæan scale for measuring plants : it is the diameter of a hair, and the twelfth part of the *Linea.* See *Mensura.*

CAPI-

CAPITATUS *Flos*, as in the *Mentha piperita, aquatica*, & *Thymus serpyllum*. See *Capitulum*.

CAPITULUM [dim. à *caput*, a head] a species of inflorescence, in which the flowers are firmly connected on the summit of the pedunculus, so as to form a kind of knob or head, as in the *Gomphrena*. A *Capitulum* is *subrotundum, globosum, dimidiatum, foliosum*, or *nudum*.

CAPREOLUS [dim. à *caprea*, a branch that produces tendrils]. A tendril. See *Cirrhus*. *Processus plantæ filamentosi quibus illa vicinis corporibus alligatur*.

· CAPSULA, a little chest, or casket; a hollow *Pericarpium* which naturally separates in some determinate manner. Its several members are called *Valvula, Dissepimentum, Columella, Loculamentum*.

CARINA, the keel of a boat or ship; the inferior *petalum* of a papilionaceous *corolla*.

CARINATUM *Folium* [*carina*, the keel or bottom of a ship] *si pars prona disci prominet longitudinaliter*, when the inferior disk or back of the leaf resembles the keel of a ship.

Carinatum

Carinatum Nectarium, as in the *Utricularia minor.*
Carinatus Calyx, as in the *Phalaris.*

CARIOPHYLLÆUS *Flos* [*Caryophyllus,* the clove-tree] composed of many *petala,* as it were emerging from the bottom of a tubular *calyx,* as in the *Caryophyllus, Linum.* Tournef. clafs the eighth.

CARNOSUM *Folium* [*caro,* flesh] a leaf of a fleshy substance, *quod interne pulpa repletum est,* but not of so close a texture as the *folium compactum,* nor so soft as the *pulposum*; exemplified in the *Sedum dasyphyllum.*
Carnosa Radix, as in the *Valeriana.*

CARTILAGINEUM *Folium* [*Cartilago,* a cartilage] *cujus margo cartilagine, a substantia folii diversissima, firmatur,* whose margin is strengthened by a cartilaginous rim of a substance different from the disk.

CARYOPHYLLEI [*Caryophyllus,* a pink or gillyflower] An order of plants in the *Fragmenta methodi naturalis* of Linnæus, containing these genera, viz. *Dianthus, Saponaria, Drypis, Cucubalus, Silene, Lychnis, Coronaria, Agrostema, Frankenia, Alsine, Cerastium, Holosteum, Arenaria, Spergula, Sagina, Moerhingia.*

D

CATENULATA *Scabrities* [*Catena*, a chain] a species of glandular *Scabrities*, hardly visible to the naked eye, resembling little chains, on the surface of some plants.

CATULUS. See *Amentum.*

CAUDEX [dict. à *cædo*] the stem of a tree; according to Linnæus, it is the ascending and descending body of the *radix*. The *Caudex ascendens* rises gradually above the surface of the earth, serving often as a trunk, and producing the herb or plant. The *Caudex descendens* strikes gradually into the ground, and spreads into *radiculæ*.

CAULESCENS *Planta* [*Caulis*] having a stem, opposed to *Acaulis.*

Caulescens radix : Ludwig. The same with the *fusiformis* of Linnæus, exemplified in the *Daucus* & *Scorzonera*. Linnæus also applies this term to the roots of the *Brassica oleracea, rapa*, & *napus.*

CAULINA *Folia* [*Caulis*, a stem] leaves growing immediately upon the stem, without the intervention of branches, as in the *Agrimonia eupatoria.*

Caulinis

Caulinis Pedunculus, the foot-stalk of a flower proceeding from the stem.

CAULIS [à *κανλὸς,* a stalk] a stem ; that species of *Truncus* common to most plants ; defined by Linnæus to be the proper trunk of the herb, which elevates the leaves and fructification.

CERNUUS *Pedunculus, Flos* [à *cerno,* to discern, *quod terram cernat*] bent, drooping, hanging down its head, *cum apice incurvatur ut flos versus latus alterum vel terram nutet, nec poterit erectus attolli ob curvaturam strictam pedunculi, uti in Carpesio, Bidente radiata, Carduo nutante, Scabiosa alpina,* &c.

CESPITOSA *Planta* [*Cespes,* turf, or sod] are those plants which produce many stems from one root, and thence form a close thick carpet on the surface of the earth.
Cespitosæ Paludes, Turf-bogs.

CILIATUM *Folium* [*Cilium,* the eye-lash] *cujus margo setis parallelis longitudinaliter obvallatur,* whose margin is guarded by parallel bristles longitudinally, as in the *Erica tetralix, ciliaris.*
Ciliata Spica, fringed with short, small, bracteal leaves.

Ciliata

Ciliata Corolla, as in the *Ruta, Menyanthes, Tropæolum.*

CIRCINALIA *Folia* [*circes,* a hoop, or ring] A term of foliation expreſſive of the leaves within the *gemma* being rolled ſpirally downward, *deorſum ſpiraliter involvuntur,* as in the *Filices* & *Palmæ nonnullæ.*

CIRCUMSCISSA *Capſula* [*circum,* about, & *cædo,* to cut] opening, not longitudinally, as in general, but tranſverſely like a common ſnuff-box, as in the *Anagallis.*

CIRRHIFERUS *Pedunculus* [*cirrus,* & *fero,* to bear] bearing a tendril, as in *Cardioſpermum, Vitis.*
  Cirrhiferum folium, as in the *Fumaria capreolata,* & *claviculata.*

CIRRHOSUM *Folium* [à *cirrhus,* a tuft or lock of hair] terminating in a tendril, as in the *Glorioſa, Flagellaria, Niſſolia.*

CIRRHUS, rather *Cirrus* [à κἐρας, *cornu,* a horn, *quod cirri cornuum figuram referant*] one of the *fulcra* of plants ; a claſper or tendril ; that ſpiral ſtring by which ſome plants fix themſelves to other bodies, *vinculum filiforme ſpirale quo plan-*
                                                              ta

ta alio corpori alligatur, as in the *Vitis, Banisteria, Cardiospermum, Pisum, Bigonia*. A *Cirrus* is termed *axillaris, foliaris, peticlaris, peduncularis*, according to the part from which it proceeds; it is *simplex, bifidus, trifidus, multifidus*, according to the number of its chords; *convolutus, revolutus*, according to its direction.

CLASSIS, a class, is by Linnæus defined to be an agreement of several *genera* in the parts of fructification, according to the principles of nature distinguished by art. He divides the vegetable kingdom into twenty-four classes, viz. 1. *Monandria*, 2. *Diandria*, 3. *Triandria*, 4. *Tetrandria*, 5. *Pentandria*, 6. *Hexandria*, 7. *Heptandria*, 8. *Octandria*, 9. *Enneandria*, 10. *Dodecandria*, 12. *Icosandria*, 13. *Polyandria*, 14. *Didynamia*, 15. *Tetradynamia*, 16. *Monadelphia*, 17. *Diadelphia*, 18. *Polyadelphia*, 19. *Syngenesia*, 20. *Gynandria*, 21. *Monoecia*, *Dioecia*, 23. *Polygamia*, 24. *Cryptogamia*.

CLAVATUS *Petiolus, Pedunculus* [*clavis*, a nail, or *clava*, a club] in its classical acceptation means studded with nails or spangles; but here it alludes to the *shape* of a common nail, tapering from its *basis* to the *apex*.

Clavatus Calyx, as in *Silene*.

Clavata Capsula, as in the *Papaver argemone*.

CLA-

CLAVICULA [dim. à *Clavis*, a key] a tendril: *Tournef.* See *Cirrhus*.

CLAUSA *Corolla*, clofed, fhut, as in the *Melampyrum pratenfe*, oppofed to *hians*.

COADUNATÆ [*coaduno*, to join, or gather together] An order of plants in the *Fragmenta methodi naturalis* of Linnæus, in which he has thefe *genera*, viz. *Annona, Liriodendrum, Magnotia, Uvaria, Michelia, Thea*.

COARCTATI *Rami* [*coarƈto*, to ftraiten or prefs together] forming very acute angles with each other; oppofed to *Divergentes*.
  Coarƈtata Panicula, when the *pedunculi* are fhort and ereƈt, and confequently the flowers compaƈt; oppofed to *diffufa*.
  Coarƈtatus Pedunculus, oppofed to *patulus*.

COCHLEATUM *Legumen* [à *Cochlea*, the fhell of a fnail] refembling the fhell of a fnail, as in the *Medicago*.

COLORATUM *Folium* [*Color*, colour] coloured; *i. e.* when thofe leaves, which are generally green, are of any other colour, *quod alium colorem quam viridem induit*.
  Coloratus Calyx, as in the *Bartfia*.

COLUMNELLA, a little column; the membranaceous fubftance which connects the internal partitions with the feed, in that fpecies of *pericarpium* termed *capfula*.

COLUMNIFERI [*Columna*, a pillar, & *fero* to bear] An order of plants in the *Fragmenta methodi naturalis* of Linnæus, in which are thefe genera, viz. *Camellia, Xylon, Hibifcus, Turnera, Malva, Urena, Malope*, &c. Columniferæ, mucilaginofæ, lubricantes, obtundentes, & maturantes funt. *Lin*.

COMA [Κομη, a bufh, or head of hair] a fpecies of *fulcra* compofed of remarkably large *bracteæ*, which terminate the *caulis*, as in the *Lavendula, Salvia, Corona imperialis*.

COMMUNIS *Gemma*, regards the contents of the *gemma*; containing both flower and leaves.

Communis Calyx, when it contains both *Receptaculum* and *Flofculi*, as in the *Tragopogon, Scorzonera*, and moft of the other plants in the clafs *Syngenefia* of Linnæus.

COMOSÆ [*Coma*, a head of hair]. An order of plants in the *Fragmenta methodi naturalis*

lis of Linnæus, containing thefe genera, viz.
Spiræa, Filipendula, Aruncus.

Comofa radix, a fpecies of the *tuberofa,* when
from the top of a bulbous root, immediately under the bafe of the ftem, a number of *fibrillæ* are
put forth, fo as to refemble a head of hair.
Ludwig.
Comofus Racemus, as in the *Fritillaria regia.*

COMPACTUM *Folium* [à *compingo,* to put
together] regards the fubftance of leaves, and
fignifies their pulp being of a clofe confiftent
texture.

COMPLETUS *Flos,* Vaill.   See *Auctus,* or
Calyculatus.

COMPOSITUS *Caulis.*   A compound ftem
is divided into *ramuli,* fmall branches, diminifh-
ing as they afcend : they are either *dichotomus,
fubdivifus,* or *articulatus.*
Compofitus Corymbus, formed of a number of
fmall *corymbi,* oppofed to *fimplex.*
Compofitus Flos, an aggregate flower compofed
of many *flofculi feffiles,* on a common entire *receptaculum,* with a common *perianthium,* and
whofe *antheræ,* being five in number, unite in
the form of a cylinder ; the *flofculi* are monopetalous,

petalous, and under each of them is a mono-
spermous *germen :* such are the class *Syngenesia*
of Linnæus. Compound flowers are either *li-
gulati, tubulosi,* or *radiati.*

Compositum Folium, in general, signifies a *petiolus* with more than one *foliolum* upon it, of
which there are the following species, *viz. compositum* properly so called, *articulatum, conjugatum, digitatum, pedatum, pinnatum, decompositum,
supradecompositum.*

Compositum Folium, properly so called, is a *petiolus* with one series of *foliola* and no more.

Composita Fructificatio, opposed to *simplex ;* composita ex flosculis.

Composita Umbella, having no *umbellulæ* on the
apices of the *pedunculi.*

Compositi, a numerous order of plants in the
Fragmenta methodi naturalis of Linnæus, where
he divides them into *Semiflosculosi, Capitati, Co-
rymbiferi,* and *Oppositifolii.*

COMPRESSUS *Caulis, Folium,* resembling a
cylinder compressed on opposite sides, the transverse section forming an ellipsis : when applied
to a leaf, it signifies, compressed in its lateral
margins, *quod a lateribus maginalibus oppositis com-
primitur, ut substantia folii major fiat quam discus,*
i. e. its depth, or thickness, exceeding its
breadth.

E

CONCAVUM *Folium*, a concave leaf, *cum margo folii arctior fit quam ut difcum circumfcribat, unde deprimitur difcus.*

CONCEPTACULUM, a receiver ; a *Pericarpium* of one *Valvula*, opening longitudinally, and not having the feeds faftened to it. Linnæus in his lateft works fubftitutes *Folliculus* for *Conceptaculum.*

CONDUPLICATUM *Folium* [*con*, & *duplicor*, to be doubled] A term in Foliation, fignifying that the fides of the leaf, within the *gemma*, are parallel and approach each other, as in the *Rofa, Fraxinus, Juglans, Amygdalus, Cerafus, Quercus, Fagus,* &c.

CONFERTI *Rami*, [à *confercio*, to fill, to ftuff] branches crouded fo as to leave hardly any fpace between ; oppofed to *remoti*.
Confertus Verticillus, when the flowers which form the *Verticillus* are numerous, and confequently crouded ; oppofed to *diftans*.
Conferta folia, as in the *Antirrhinum monfpeffulanum*, & *linaria*.

CONFLUENTIA *Folia* [*confluo*, to flow together] growing in tufts partially, fo as to leave the intermediate parts of the *caulis* quite bare.
CON-

CONGLOBATUS *Flos* [con, & *globus*, a ball] *Capitatus* of Ray, *Compositus* of Tournefort and Linnæus. See *Compositus*.

CONGLOMORATI *Flores* [con, & *glomus*, a clew] growing on a branching *Petiolus*, and closely, but irregularly, connected ; opposed to *Paniculæ diffusæ*.

CONGESTA *Umbella* [à *congeror*, to be heaped] the flowers closely collected into a spherical shape, as in the *Allium* ; opposed to *divaricata, divergens, distans,* &c.

CONICA *Scabrities* [Κῶνος, *conus*, a cone] a species of setaceous *Scabrities*, scarce visible to the naked eye, on the surface of some plants, in which the minute bristles are conical.

CONIFERÆ [ex Κῶνος, a cone, & *fero*, to bear] An order of plants in the *Fragmenta methodi naturalis* of Linnæus, containing these genera, viz. *Abies, Pinus, Cupressus, Thuja, Juniperus, Taxus, Ephedra*. Coniferæ sunt resiniferæ & diureticæ. *Lin.*

CONJUGATUM *Folium* [ex con, & *jugo*, to couple] that species of pinnate leaf which

has

has two *foliola*, and no more, upon each *petiolus*; *cum pinnatum conſtat modo foliolis duobus, nec pluribus.*

Conjugatus Racemus, when two *Racemi* are united by a common *pedunculus.*

CONNATUM *Folium* [ex *con,* & *naſcor,* to be born, to grow together] *ſi folia oppoſita inter ſe connata in unum,* when two oppoſite leaves unite ſo as to have the appearance of one leaf, as in the *Lonicera, Eupatorium.*

Connata Stamina, united, cleaving together.

CONNIVENS *Corolla* [*connivo,* to wink] when the *apices* of the *petala* converge, ſo as to cloſe the flower, as in the *Trollius europæus.*

Conniventes Antheræ, approaching or inclining towards each other, as is frequent in the claſs *Didynamia* of Linnæus.

CONTINUATUM *Folium,* continued; when the leaf appears to be a continuation of the ſubſtance of the *caulis,* as in ſome ſpecies of the *Ficus.* Ludw.

CONTORTI [*contorqueo,* to twiſt] An order of plants in the *Fragmenta methodi naturalis* of Linnæus, containing the following *genera,* viz.

viz. *Rauwolfia, Tevetia, Cerbera, Plumieria, Ta-bernæmontana, Cameraria, Nerium, Vinca, Apo-cynum, Cynanchum, Creopegia, Afcelpias, Stapelia.*

Contorti, a natural clafs in Scopuli's *Flora Carniolica.*

CONTRARIÆ *Valvulæ.* The valves are termed *contrariæ,* when the *diffepimentum* is placed tranfverfely between them ; oppofed to *paralleli* ; exemplified in the *Subularia.*

CONVEXUM *Folium,* a convex leaf; *i. e.* rifing from the margin to the centre of the difk : *quod in difco magis elevatum eft* ; the reverfe of *concavum.*

CONVOLUTUS *Cirrus* [*convolvo,* to wrap round] a tendril twining in the fame direction with the fun's motion, that is, from eaft, by the fouth, to the weft : oppofed to *Revolutus.*

Convolutum Folium, a term of Foliation, fig-nifying the leaf being rolled up like a fcroll of paper ; *unius lateris margo circumambit alterum ejuf-dem folii marginem inftar cuculli* ; as in the *Arum, Piper, Solidago, Braffica, Prunus,* & *Gramina pleraque.*

CONUS. See *Strobilus.*

COR-

CORCULUM [dim. à *cor*, the heart] the heart and effence of a feed, and the *primordium* of the future plant, attached to, and involved in the *Cotyledon :* it confifts of the *Plumula*, which Linnæus calls its effence, and *Roftellum*, which fee.

CORDATUM *Folium* [*Cor*, the heart] *eft o-vatum bafi excavatum, deftitutum angulis pofticis ;* fhaped like the heart on cards, but the *apex* not quite fo fharp ; as in the *Potamogeton perfoliatum, Menyanthes nymphoides, Menifpermum virginic. carolin. & cocculus.*

CORDIFORMIS. See *Cordatum*.

COROLLA [dim. à *corona*, a crown] one of the feven parts of fructification, according to Linnæus, who defines it thus, *liber plantæ in flore præfens*, the bark of the plant prefent in the flower. It is the coloured or painted leaves of the plant, confifting of *petala* and *nectarium*.

COROLLULA [dim. à *corolla*] a little *corolla*.

CORONA *Seminis*, a crown ; the little crown which adheres to many kinds of feeds, and which, ferving them as wings, enables them to difperfe :

difperfe : it is either *Calyculus* formed of the *perianthium* of the flower, as in the *Scabiofa, Knautica, Ageratum, Arctotis* ; or it is *Pappus*, as in the *Hieracium, Sonchus, Crepis, Scorzonera, Tragopogon.*

CORONARIÆ [*Corona*, a crown] An order of plants in the *Fragmenta methodi naturalis* of Linnæus, in which are the following genera ; viz. *Ornithogalum, Scilla, Hyacinthus, Afphodelus, Anthericum, Polyanthes.*

CORONULA, dim. à *Corona*, which fee.

CORTEX [à *corium*, a hide, & *tego*, to cover] the outer rind of vegetables diftinct from the *liber :* thus the *corolla* is a continuation of the *liber*, and the *calyx* of the *cortex.*

CORTICALIS *Gemmatio* [*cortex*, rind or bark] regards the origin of the *gemma* ; proceeding from the *cortex* of the plant ; oppofed to *petiolaris, ftipularis.*

CORYDALES [an à Κορυς, *galea, caffis, galerita*] An order of plants in the *Fragmenta methodi naturalis* of Linnæus, containing the following genera, viz. *Melianthus, Epimedium, Hypecoum.*

pecoum, *Fumaria, Impatiens, Leontice, Monotro-*
pa, Utricularia, Tropæolum.

CORYMBUS, in its proper acceptation, is
a clufter of ivy-berries. Linnæus makes it a
fpecies of inflorefcence, in which the flowers
grow in clufters, each upon a feparate *peduncu-*
lus, as in the filiquofe plants in general, viz.
Myagrum, Anaftatica, Cochlearia, &c. *Corymbus*
fit ex fpica, dum finguli flores pedunculis propriis
inftruuntur, fitu elevato proportionali. The *Inflo-*
refcentia corymbifera may be *fimplex* or *compofita.*

COTYLEDON [Κόϑυλη, *cavitas,* a cavity] the
lateral, bibulous, perifhable lobe of the feed.

CRENATUM *Folium* [*Crena,* a notch] *cujus*
margo angulis, neutram extremitatem refpicientibus,
fecatur, whofe margin is cut with fmall teeth,
or angles, inclining towards neither extremity,
fo that *radii* drawn from the centre of the
leaf would accurately bifect each faliant angle.
When inftead of angles the margin is formed
of a number of fegments of fmall circles, it is
then termed *obtufe crenatum*; when the larger
notches are themfelves furrounded by fmall
ones, *duplicate crenatum.* The *folium crenatum* is
exemplified in the *Primula farinofa.*
  Crenata Corolla, as in *Linum.*

CRI-

CRISPUM *Folium*, undulated from the margin of the leaf being too long for the difk, *cum peripheria folii major evadit quam difcus admittit, ut undulatum fiat.* *Folia crifpa* are always monftrous productions.

CRISTATUS *flos* [à *crifta*, a tuft or creft.] crefted, as in the *Polygala*.

CRUCIFORMES *Flores* [*Crux*, a crofs, & *forma*] crofs-fhaped; confifting of four *petala* regularly difpofed in the form of a crofs: they conftitute the fifth clafs in Tournefort, and the *Tetradynamia* of Linnæus.

CRYPTANTHERÆ [à κρυπτω, *occulto*, to hide, & ανθος, *flos*, a flower] the nineteenth clafs in Royen's fyftem, comprehending thofe plants whofe fructification is concealed, *viz.* part of the *Filices*, *Mufci*, *Algæ*, & *Fungi*.

CRYPTOGAMIA [Κρυπτος, *occultus*, concealed, & Γαμος, *nuptiæ*, nuptials] the twenty-fourth clafs in the Linnæan fyftem, comprehending thofe plants whofe fructification is concealed, either through minutenefs, or within the fruit: *nuptiæ clam celebrantur.* The orders are four, *viz.* FILICES, containing 16 *genera*, *viz. Equifetum, Onoclea, Ophiogloffum, Ofmunda,*

Achroftichum,

Achrostichum, Polypodium, Hemionitis, Asplenium, Blechnum, Lonchitis, Pteris, Adianthum, Tricho- manes, Marsilea, Pilularia, Isoetes ; Musci, containing 11 *genera,* viz. *Lycopidium, Porella, Spagnum, Phascum, Splachnum, Polytrichum, Mni- um, Bryum, Hypnum, Fontinalis, Buxbaumia* ; Alg.e, containing 14 *genera,* viz. *Marchan- tia, Jungermannia, Targionia, Anthocerus, Brasia, Riccia, Lichen, Byssus, Tremella, Ulva, Fucus, Conferva, Spongia, Chara* ; Fungi, contain- ing 10 *genera,* viz. *Agaricus, Boletus, Hydnum, Phallus, Clathrus, Elvela, Peziza, Clavaria, Ly- choperdon, Mucor.* Cryptogamia vegetabilia fæ- pius fufpecta continet. *Lin.*

CUBITUS [à *cubando,* lying down, *quod ad fumendos cibos in ipfa cubamus*] a cubit. The ninth degree in the Linnæan fcale for meafu- ring plants : from the elbow to the extremity of the middle finger ; or feventeen Parifian inches.

CUCULLATUM *Folium* [*cucullus,* a coronet of paper in which grocers put their fpices] roll- ed up like a cone, longitudinally, as in the *Ge- ramum Afric.* It alfo fignifies hooded, cowled as in the leaves of the *Sarracenia.*

CUCURBITACEÆ [*Cucurbita,* a gourd]
'An

An order of plants in the *Fragmenta methodi naturalis* of Linnæus, which are thefe *genera*, viz. *Paffiflora, Fevillea, Momordica, Trichofanthus, Cucumis, Cucurbita, Bryonia, Sicyos, Melothria, Gronovia.*

CULMINIÆ [*culmen*, the top or crown of any thing] An order of plants in the *Fragmenta methodi naturalis* of Linnæus, in which are thefe *genera*, viz. *Tilia, Theobroma, Sloanea, Bixa, Heliocarpus, Triumphetta, Bartramia,* &c.

CULMUS [ex κάλαμος, *calamus*, a reed or ftraw] that fpecies of *Truncus* proper to graffes; it elevates the leaves and the fructification.

CUNEIFORME *Folium* [*Cuneus*, a wedge] fhaped like a wedge, *fenfim deorfum anguftatur*, its apex next the ftem, as in the *Apium graveolens, Saxifraga trydactylites.*

CUSPIDATUM *Folium* [*Cufpis*, the point of a fpear]. This term regards the *apex* only, and is applied to thofe leaves whofe *apex* refembles the point of a lance or fpit.

CYATHIFORMIS *Corolla* [*Cyathus*, a cup] partly cylindrical, but growing wider towards the top.

CYLINDRACEA *Spica* [*cylindrus*, a roller, a cylinder] cylindrical ; equal in diameter from top to bottom.

CYLINDRICA *Scabrities* [*cylindrus*] A fpecies of *Scabrities* of a cylindrical form, on the furface of fome plants, fcarce vifible to the naked eye.

Cylindricus Calyx, as in the *Euphrafia*.

CYMA [Κῦμα, *fœtus*] a *Receptaculum* proceeding from an univerfal centre, running out into *Pedunculi faftigiati*, but with irregular partial *pedunculi*.

CYMOSUS *Flos* [*Cyma*, a fprout] an aggregate flower, whofe *Receptaculum* is divided into primary faftigiate *Pedunculi* proceeding from an univerfal centre ; but whofe fecondary *Pedunculi* are irregular, which diftinguifhes it from the *Umbella*.

Cymofæ, an order of plants in the *Fragmenta methodi naturalis* of Linnæus, containing thefe *genera*, viz. *Diervilla, Lonicera, Mitchella, Loranthus, Ixora, Morinda, Cinchona.*

CYTINIFORMIS *Calyx* [à *Cytinus*, the flower of the pomegranate] refembling the *Calyx*

lyx of the pomegranate, i. e. *campaniformis, multifidus.* Tournef.

D.

DÆDALEUM *Folium* [Δαιδαλ⊙, *dædalus,* ingenious] a leaf whose texture or shape is remarkably beautiful and exquisitely wrought. Linnæus has not, I think, any where explained this term, but it seems to admit of no other meaning.

DEBILIS *Caulis* [ex *de* & *habilis*] weak, feeble, as in the *Veronica chamædrys.*

DECAGYNIA [δεκα, *decem,* ten, & Γυπ, *mulier,* a woman] The fifth order in the tenth class in the Linnæan system; comprehending those plants whose fructification discovers *ten Styli,* which are considered as the female organs of generation.

DECANDRIA [δεκα, *decem,* ten, & ανης, *maritus,* a husband]. Linnæus's tenth class, comprehending those hermaphrodite plants which bear flowers with ten *stamina.* This class hath five orders, *viz.* Monogynia, including 45 *genera, viz. Sophora, Anagyris, Cercis, Bauhinia,*

Bauhinia, Parkinsonia, Hymenæa, Cassia, Poinciana, Cæsalpinia, Guilandina, Guajacum, Cynometra, Anacardium, Swietenia, Dictamnus, Ruta, Toluifera, Hæmatoxylum, Adenanthera, Melia, Trichilia, Zygophyllum, Quassia, Fagonia, Tribulus, Thryallis, Limonia, Monotropa, Jussiæa, Quisqualis, Dais, Bucida, Copaifera, Samyda, Melastoma, Kalmia, Ledum, Rhodora, Rhododendron, Andromeda, Epigæa, Gaultheria, Arbutus, Clethra, Pyrola ; DIGYNIA containing 11 *genera,* viz. *Royena, Hydrangea, Cunonia, Chrysosplenium, Saxifraga, Tiarella, Mitella, Scleranthus, Gypsophila, Saponaria, Dianthus* ; TRIGYNIA, containing 10 *genera,* viz. *Cucubalus, Silene, Stellaria, Arenaria, Cherleria, Garidella, Malpighia, Banisteria, Triopteris, Erythroxylon* ; PENTAGYNIA, containing 11 *genera,* viz. *Averrhoa, Spondias, Cotyledon, Sedum, Penthorum, Oxalis, Suriana, Agrostemma, Lychnis, Cerastium, Spergula* ; DECAGYNIA, containing two *genera,* viz. *Neurada, Phytolacca.*

DECAPHYLLUS *Calyx* [δέκα, *decem,* ten, & Φυλλον, *folium,* a leaf] consisting of ten leaves, as in the *Hibiscus.*

DECIDUUM *Folium* [*decido,* to fall down, to die] a term expressing the second degree of duration ;

duration; fee *Caducum:* falling off with the flower.

Deciduæ Stipulæ, as in the *Padus, Cerafus, Amygdalus; Populus, Tilia, Ulmus, Quercus, Fagus, Alnus, Ficus, Morus.*

Deciduus Calyx, as in the *Berberis,* and the clafs *Tetradynamia.*

DECLINATUS *Caulis,* declined. The firft and leaft degree of curvature towards the earth. See *Incurvatus* and *Nutans.*

DECOMPOSITA *Folia,* are thofe compound leaves which confift of many *foliola* on a once-divided *petiolus, cum petiolus femel divifus adnectit foliola plura.* See *Compofitum folium.* The different kinds of *Folia decompofita,* are *bigemina, biternata, bipinnata,* which fee.

DECUMBENS *flos* [*decumbo,* to lie down] drooping, as in *Caffia, Diadelphia omni.*

DECURRENS *Folium* [*decurro,* to run along] when the *bafis* of a feffile leaf extends downward along the *caulis,* below the proper termination of the leaf; *i. e.* when the *bafis* is long in proportion to the leaf, and adheres entirely to the ftem, as in the *Verbefina, Carduus, Sphæranthus, Verbafcum thapfus.*

DE-

DECURSIVE *Folium pinnatum* [*decurro*, to run along] when the bases of the *foliola* are continued along the sides of the *petiolus*, as they are along the stem in the *folium decurrens*.

DECUSSATA *Folia* [*decusso*, to divide] when the leaves grow in hairs and opposite, each hair being alternately on opposite sides of the stem ; *ut si planta verticaliter inspiciatur, folia quadrifariam vergunt.*

DEFLEXUS *Ramus* [*deflecto*, to bow or bend] a little bent; the least degree of curvature. See *Reflexus, Retroflexus.*

DEFLORATIA *Stamina*, having shed, or discharged, their *farina fecundens*. Stachys, stamina *deflorata* versus latera reflexa.

DEFOLIATIO [*de* & *folium*] comprehends the precise time in autumn when a plant sheds its leaves.

DELTOIDES *Folium* [Δ, *delta*, the Greek D]. Certainly this Δ has but three angles, and yet Linnæus insists upon his *folium deltoides* having four, *rhombeum ex quatuor angulis, e quibus laterales minus a basi distant quam reliqui*, the two lateral angles nearer the basis than the other two.

two. Now the figure, in the *Phil. Botanica*, which is intended to illuſtrate this explanation is like nothing in nature, and of a form quite different from thoſe leaves to which this term is applied in the *Syſtema Naturæ*, &c. as for inſtance, in the *Populus nigra,* which has four angles, and the two lateral ones are nearer the baſe than the *apex*. In order to ſolve theſe difficulties, let us firſt recollect that the *delta* was the figure of four, of the Greeks, and that 4-angular would expreſs a figure with four angles, though the figure 4 contains but three. But if this does not ſatisfy us, let us recollect that Δελτⷫ, *pugillares*, were a kind of writing-tablets, which, though triangular when ſhut, muſt neceſſarily, when open, be quadrangular. Dr Hill, in his explanation of this term, by way of example, inſtances the Sea Purſlane, which happens not to have a deltoide leaf, but *Obovatum*. Linnæus applies the term *Deltoides* to the leaves of the following plants, viz. *Populus nigra, Atriplex laciniata, Atriplex haſtata, Atriplex patula, Chenopodium ſerotinum.*

DEMERSUM *Folium* [*demergo*, to dive] in aquatic plants, ſunk below the ſurface of the water. The ſame as *Submerſum*.

DENDROIDIS *ſurculus* [à δενδρον, *arbuſtum*, a
G   ſhrub]

fhrub] fhrubby ; a fubdivifion of the *Surculus* in the *genus Hypnum.*

DENTATUM *Folium* [*Dens,* a tooth] according to Linnæus, *quod acumina horizontalia, folii confiftentia, fpatio remota habet* ; i. e. having horizontal points, of the fame confiftence with the leaf, at a little diftance from each other. If, inftead of horizontal, he had wrote, in the plane of the difk of the leaf, it would have been more intelligible ; exemplified in the *Leontodon, Primula vulgaris,* & *veris, Epilobium montanum.*

DENTICULATA *Semina* [*denticulus,* a little tooth] as in the *Bidens.*
Denticulatum Folium, as in the *Hefperis matronalis.*

DENUDATÆ [*denudor,* to be ftripped naked] An order of plants in the *Fragmenta methodi naturalis* of Linnæus, comprehending thefe genera, viz. *Crocus, Gethyllis, Bulbocodium, Colchicum.*

DEPENDENS *Folium* [*dependeo,* to hang down] *quod refta terram fpeftat,* pointing directly to the ground.

DETERMINATIO *Foliorum.* By the determination

termination of leaves is meant fome particularity exclufive of their proper ftructure, *viz.* place, number, fituation, direction, and infertion.

DIADELPHIA [Δίς, *bis*, twice, & Αδελφος, *frater*, a brother] Clafs the feventeenth in the fexual fyftem, comprehending thofe plants which bear hermaphrodite flowers with two fets of united *Stamina*; but this circumftance muft not be abfolutely depended on. They are the *Papilionacei* of Tournefort, the *Irregulares tetrapetali* of Rivinus, and the *Leguminofæ* of Ray. The flowers are generally pendulous. This is the moft natural of all the claffes. The claffical characteriftics are thefe, *viz.* CALYX, *Perianthium*, monophyllous, campanulate, withering; *Bafis* gibbous, attached below to the *pedunculus*, and obtufe at the top; the brim, quinquedentate, acute, erect, oblique, unequal, the lower odd denticle longeft, and the upper pair fhorteft, and farther afunder; the bottom moift with a melleous liquor, inclofing the *receptaculum*. The *Calyx*, being of confequence in determining the *genera*, merits particular attention. COROLLA, papilionaceous, unequal, whofe *petala* are diftinguifhed in the following manner; *Vexillum*, a petal covering the reft, incumbent, larger, plano-horizontal, its *unguis* inferted into the fuperior margin of the *receptaculum,*

culum, roundish on the outside of the *calyx,* almost entire, with a longitudinal ridge especially towards the *apex,* the part of the petal nearest the basis almost semicylindrical, embracing the parts beneath, the *discus* of the *petalum* depressed on both sides, turning upwards near the margin ; where the divided tube ends, and the limb begins to unfold, are two cavities, which compress the *alæ* beneath : *Alæ,* two *petala,* equal, on each side the flower, under the *vexillum,* their margins incumbent, parallel, subrotundo-oblong, broader outwards, the upper margin straighter, the lower rounder ; the *basis* of each bifid, the inferior part stretched into an *unguis,* inserted into the side of the *receptaculum,* about the length of the *calyx,* the upper shorter, and bent : *Carina,* the lowest *petalum* often bipartite, under the *Vexillum,* and between the *alæ,* boat-shaped, concave, the sides compressed, set like a boat in the water, the *basis* mutilate, of which the inferior part extends into an *unguis* the length of the *calyx,* and inserted into the *receptaculum* ; but the lateral and superior short *laciniæ* are infolded with the correspondent part of the *alæ* ; the sides of the *carina* are similar to the *alæ* both in shape and situation, except that they are inferior and interior ; the carinal line runs straight as far as the middle, and then rises in the segment of a circle,

but

but the marginal line runs ſtraight to the *apex,* where the two lines meet, and terminate obtuſely. STAMINA, *diadelphia,* two *filamenta* of different forms, the one inferior involving the *piſtillum,* the other ſuperior on the *piſtillum* incumbent ; the inferior filament involving the *germen,* membranaceous from the middle downwards, cylindrical, opening longitudinally upwards, the upper part terminating in nine ſubulate *radii,* i-mitating the *corolla* in flexure and length, the lower *radii* being gradually longer, the ſuperior filament ſubulato-ſetoſe, covering the fiſſure of the other, incumbent on it, alike in ſituation, gradually ſhorter, ſimple, its *baſis* detached from the other, affording a vent for the honey on each ſide ; *Antheræ,* ten in all, one on the ſuperior filament, and one on each of the nine *radii* of the inferior. PISTILLUM, ſingle, grow · ing out of the *receptaculum* within the *calyx* ; *Germen,* oblong, a little compreſſed, ſtraight, of the length of the inferior filament by which it is involved ; *Stylus,* ſubulato-filiform, a-ſcending, of the ſame length and poſition with the *radii* of the filament among which it ſtands, withering : *Stigma,* downy, of the length of the *Stylus* from the part turned upwards, and placed immediately under the *antheræ.* PERI-CARPIUM, *Legumen* oblong, compreſſed, obtuſe, bivalvate, with a longitudinal ſuture both above

and

and below, both ſtraight, yet the upper one de-
ſcends near the baſis, and the lower riſes near
the *apex*, opening at the upper ſuture. SEMI-
NA few, roundiſh, ſmooth, fleſhy, pendulous,
prominent with an embryo towards the point of
inſertion ; the *ova* being diſcharged, the Cotyledons retain the forms of the divided ſeed :
RECEPTACULUM, the proper *receptacula* of the
ſeeds are very ſmall, very ſhort, thinner at the
baſis, obtuſe at the diſk, oblong, inſerted longitudinally and alternately in the upper ſuture
of the *Legumen*. The orders are 4, viz. PENTAN-
DRIA, containing but one *genus*, viz. *Monieria* ;
HEXANDRIA, containing but one *genus*, viz.
Fumaria ; OCTANDRIA, containing two *genera*,
viz. *Polygala & Securidaca* ; DECANDRIA, containing 44 *genera*, viz. *Amorpha, Ebenus, Erythrina, Spartium, Geniſta, Lupinus, Anthyllis,
Æſchynomene, Piſcidia, Borbonia, Aſpalathus, O-
nonis, Crotalaria, Colutea, Phaſeolus, Dolichos,
Orobus, Piſum, Lathyrus, Vicia, Aſtragalus, Bi-
ſerrula, Phaca, Pſoralea, Trifolium, Glycyrrhiza,
Hedyſarum, Coronilla, Ornithopus, Scorpiurus,
Hippocrepis, Medicago, Trigonella, Glycine, Clitoria,
Robinia, Indigofera, Ulex, Cicer, Ervum, Cy-
tiſus, Galega, Lotus, Arachis.* Papilionaceorum
folia jumentis & pecoribus, ſemina variis animalibus eſculenta ; ſunt farinacea & flatulenta.
Lin.

DIANDRIA [Δὶς, *bis*, twice, & *ἀνηρ*, *mari-tus*, a hufband] the fecond clafs in the Linnæan fyftem, comprehending fuch hermaphrodite flowers as have two *ftamina*: it includes three orders, *viz.* MONAGYNIA, in which are 26 *ge-nera*, viz. *Nictanthes, Jafminum, Liguftrum, Phil-lyrea, Olea, Chionanthus, Syringa, Eranthemum, Circæa, Veronica, Pæderota, Jufticia, Dianthe-ra, Gratiola, Pinguicula, Utricularia, Verbena, Lycopus, Ametyftea, Cunila, Ziziphora, Monar-da, Rofmarinus, Salvia, Collinfonia, Morina*; DIGYNIA, containing one *genus,* viz. *Antho-xanthum*; TRIGYNIA, of which alfo there is but one genus, viz. *Piper.*

DIANGIÆ [δὶς, & *αγγ⊕*, *vas,* a veffel, or *loculamentum*] the fixteenth clafs in Boerhaave's fyftem, containing *Lythrum, Saxifrage, &c.*

DICHOTOMUS *Caulis* [Διχοτομος, *diffectus,* divided] forked, *bifariam femper divifus.*

Dichotomus Pedunculus, as in the *Meliffa cala-mintha.*

DICOTYLEDONES [δὶς, & *cotyl.*] A mode of placentation, fignifying that the feeds have two cotyledons: thefe are either *immutatæ, pli-catæ, duplicatæ, obvolutæ, fpirales,* or *reducta.*

DIDYMA *Anthera* [Διδυμⓢ, *geminus*, twins] two upon each *filamentum*, as in the *Ranunculus*.

DIDYNAMIA [Δις, *bis*, twice, & Δύναμις, *potentia*, power] Linnæus's fourteenth clafs, comprehending thofe plants which produce hermaphrodite flowers bearing one *piſtillum*, and four *ſtamina*, two of which are long and two fhort. Its claffical characteriſtics are thefe : CALYX, *Perianthium*, monophyllous, erect, tubulated, quinquefid, generally with unequal fegments, perfifting. COROLLA, monopetalous, erect, whofe *baſis* is tubulated, and ferves the purpofe of a *nectarium:* the *limbus* is generally ringent, its fuperior *labium* upright, the inferior extended horizontally, trifid, the broadeſt lobe in the middle. STAMINA, whofe filaments are fubulate, inferted into the tube of the *Corolla*, and inclining towards its back : the innermoſt are the fhorteſt ; they are all parallel, and feldom exceed the length of the *Corolla*. The *Antheræ* are generally hid under the fuperior *labium* of the *Corolla*, in pairs, refpectively connivent. PISTILLUM, the *Germen* commonly above the *Receptaculum*, a fingle filiform *Stylus*, bent with the *filamenta*, and generally inclofed within them, their fummits a little curved, and the *ſtigma* emarginate. PERICARPIUM, either entirely wanting, or generally bilocular.

SEEDS,

SEEDS, if there be no *Pericarpium*, are four, lodged in the bottom of the *Calyx*; if there be a *Pericarpium*, they are more numerous, and are fixed to a *Receptaculum* in the middle of it. Those of the first order are the *Labiati*, and of the second, the *perſonati* of Tournefort. The orders two, viz. GYMNOSPERMIA, which, except *Phryma*, have univerſally four ſeeds; the *ſtigma* bipartite and acute, with the inferior *lacinia* reflexed; it contains 33 *genera*, viz. *Adjugo*, *Teucrium*, *Satureja*, *Thymbra*, *Hyſſopus*, *Nepeta*, *Lavandula*, *Betonica*, *Sideritis*, *Mentha*, *Glecoma*, *Orvala*, *Lamium*, *Stachys*, *Galeopſis*, *Ballota*, *Marrubium*, *Leonurus*, *Phlomis*, *Moluccella*, *Clynopodium*, *Origanum*, *Thymus*, *Meliſſa*, *Dracocephalum*, *Horminum*, *Melittis*, *Ocymum*, *Trichoſtema*, *Scutellaria*, *Prunella*, *Praſium*, *Phryma*; ANGIOSPERMA, containing 59 *genera*, viz. *Bartſia*, *Rhinanthus*, *Euphraſia*, *Melampyrum*, *Lathræa*, *Schwalbea*, *Tozzia*, *Pedicularis*, *Gerardia*, *Chelone*, *Geſneria*, *Antirrhinum*, *Cymbaria*, *Craniolaria*, *Martynia*, *Torenia*, *Beſleria*, *Scrophularia*, *Celſia*, *Digitalis*, *Bignonia*, *Citharexylum*, *Halleria*, *Creſcentia*, *Gmelina*, *Petrea*, *Lantana*, *Cornutia*, *Loeſelia*, *Capraria*, *Selago*, *Hebenſtretia*, *Erinus*, *Buchnera*, *Browallia*, *Linnæa*, *Sibthorpia*, *Limoſella*, *Stemodia*, *Æginetia*, *Obolaria*, *Orobanche*, *Dodartia*, *Lippia*, *Seſamum*, *Mimulus*, *Ruellia*, *Barleria*, *Duranta*, *Ovieda*, *Elliſia*, *Volkameria*, *Clerodendrum*,

H

rodendrum, *Vitex, Bontia, Columnea, Acanthus, Pedalium, Melianthus.*

DIFFORMIA *Folia* [δις, & *forma,* form, fhape] when on the fame tree there are leaves of different forms; *folia diverſæ figuræ*; as in the *Tithymalus heterophyllus, Rudbeckia foliis inferioribus trilobis, superioribus indiviſis; Hibiſcus, foliis inferioribus integris, superioribus trilobis; Lepidium, foliis caulinis pinnato-multifidis, ramis cordatis amplexicaulibus integris.*

Difformis Flos, Anomalus of Tournefort, *Irregularis* of Linnæus, which fee.

DIFFUSUS *Caulis:* diffufed; *ramis patentibus*; with fpreading branches, as in the *Teucrium ſcordium.*

Diffuſa Panicula, when the *pedunculi* are long, and not very near each other, the flowers being confequently difperfed; oppofed to *coarctata.*

DIGITATUM *Folium* [*digitus,* a finger] that fpecies of compound leaf in which more than one *foliolum* is connected on the extremity of one *petiolus:* they are termed *binatum, ternatum, quinatum,* according to the number of *foliola* of which they confift.

DIGYNIA [δις, & Γυνη, *mulier,* a woman] The

The second order in each of the firſt thirteen claſſes, except the ninth, in the Linnæan ſyſtem ; it comprehends thoſe plants in whoſe fructification there are *two Piſtilla*, which are conſidered as the female parts of generation.

DIMIDIATUM *Capitulum*, [*dimidius*, half] hemiſpherical, reſembling half a head.

DIOECIA [δις, *bis*, & Οιϰϑ-, *domus*, a houſe] The twenty-ſecond claſs in the Linnæan ſyſtem, conſiſting of thoſe plants which, having no hermaphrodite flowers, produce male and female flowers on ſeparate plants ; *mares et feminæ habitant in diverſis thalamis & domiciliis.* In every ſpecies of this claſs there is both a male and a female plant diſtinctly. The males are produced from the female ſeeds, which however require the vicinity of a male plant, without which they do not propagate. It is neceſſary to obſerve, with regard to the diſtinguiſhing character of this claſs, that there are ſome particular plants excluded, notwithſtanding that they produce male and female flowers on ſeparate plants, as in the *Morus, Urtica, Croton, Rumex, Silene, Carex, Rhus, Laurus, Valeriana, Rhamnus, Cucubalus,* &c. but it does not run uniformly through the whole *genus* in any of them. The orders are fourteen, *viz.* Mo-

NANDRIA,

NANDRIA, containing but one *genus*, viz. *Na-jas*; DIANDRIA, containing 3 *genera*, viz. *Va-lifneria*, *Cecropia*, *Salix*; TRIANDRIA, contain-ing 3 *genera*, viz. *Empetrum*, *Ofyris*, *Excoecaria*; TETRANDRIA, containing 5 *genera*, viz. *Hip-pophæ*, *Trophis*, *Vifcum*, *Batis*, *Myrica*; PEN-TANDRIA, containing 11 *genera*, viz. *Ceratonia*, *Irefine*, *Cannabis*, *Humulus*, *Piftacia*, *Zanonia*, *Spinacia*, *Acnida*, *Antedefma*, *Zanthoxylon*, *Fewil-lea*; HEXANDRIA, containing 4 *genera*, viz. *Smilax*, *Tamus*, *Diofcorea*, *Rajania*; OCTAN-DRIA, containing 2 *genera*, viz. *Populus*, *Rhodio-la*; ENNEANDRIA, containing 2 *genera*, viz. *Mercurialis*, *Hydrocharis*; DECANDRIA, con-taining 4 *genera*, viz. *Datifca*, *Carica*, *Kiggelaria*, *Coriaria*; POLYANDRIA, containing but one *ge-nus*, viz. *Cliffortia*; MONADELPHIA, containing 5 *genera*, viz. *Juniperus*, *Taxus*, *Ephedra*, *Ciffam-pelos*, *Adelia*; SYNGENESIA, containing but one *genus*, viz. *Rufcus*; GYNANDRIA, containing but one *genus*, viz. *Clutia*.

DIPETALA *Corolla* [δις, & Πεταλον, *peta-lum*] confifting of two *petala*, as in the *Circæa*, *Commelina*.

DIPHYLLUS *Calyx* [δις, & Φυλλον, *folium*, a leaf] confifting of two leaves, as in the *Papaver*, *Fumaria*.

DI-

DIPLOSANTHERÆ [Δ:πλο☉, *duplex*, double, & *Anthera*] The seventeenth class in Royen's system, containing those plants whose *Antheræ* exceed the number of *petala* or segments of the *Corolla*, as far as double the number. It includes most of Linnæus's *Octandria, Decandria, & Dodecandria.*

DEPRESSUM *Folium*, depressed, *quod in disco magis deprimitur, quam ad latera*; hollow in the centre; the reverse of *gibbum*, and very different from *compressum*, which see.

Depressa Radix, its transverse diameter being greater than the longitudinal, as in the *Brassica rapa.*

DISCUS, a disk, or quoit; the middle part of a radiate compound flower, which is surrounded by the *Radius*. When applied to a leaf, it means the whole surface, circumscribed by the margin; and is either *supinus*, superior, or *pronus*, inferior.

DISPERMA [δις, & *Sperma*, a seed] producing two seeds, as the *Umbellatæ & Stellatæ.*

DISSECTUM *Folium*. See *Laciniatum.*

DISSEPIMENTUM, a partition; *paries quo fructus*

fructus interne diftinguitur in concamerationes plures, the partition which diftinguifhes the cavity of *Pericarpia* into feparate cells.

DISSILIENS *Siliqua* [*d'ffilio*, to break, to fhi-ver] burfting with clafticity, as in the *Dentaria* & *Cardamine.*

DISTANS *Verticillus,* diftant : when the flowers which compofe the *Verticillus*, being few in number, are diftant from each other.
Diftantia Stamina, as in the *Mentha.*

DISTICHA *Folia, Rami* [ex δις, *bis*, & Στιχος, *ordo*, rank] growing in two rows, or lines drawn from the *bafis* to the *apex* of the ftem or branch, as the leaves of the *Abies* & *Diervilla.*
Difticha Spica, the flowers growing in two lines, *ad utrumque latus fpectantibus*, oppofed to *Secunda.*

DIVARICATI *Rami* [*divarico*, to ftride] branches ftraddling wide from each other.

DIVERGENTES *Rami* ; diverging, oppo-pofed to *coarctati.*

DODECANDRIA [δωδεκα, *duodecim*, twelve, & ανηρ, *maritus*, a hufband] comprehends thofe
plants

plants which produce hermaphrodite flowers, which have from twelve to nineteen *Stamina*, both numbers inclusive. In this class there are five orders, *viz*. MONOGYNIA, which includes 16 *genera*, viz. *Asarum, Gethillis, Bocconia, Rhisophora, Blakea, Garcinia, Styrax, Winterania, Halesia, Crateva, Triumfetta, Peganum, Nitraria, Portulaca, Lythrum, Ginora* ;. DIGYNIA, containing 2 *genera*, viz *Heliocarpus, Agrimonia* ; TRIGYNIA, containing 2 *genera*, viz. *Reseda, Euphorbia* ; PENTAGYNIA, containing but one *genus*, viz. *Glinus* ; OCTAGYNIA, containing but one *genus*, viz. *Illicium* ; DODECAGYNIA, containing but one *genus*, viz. *Sempervium*.

DODRANS. The seventh degree in the Linnæan scale for measuring the parts of plants : the space between the extremity of the thumb and that of the little-finger when both extended ; or nine Parisian inches. See *Mensura*.

DODRANTALIS *Caulis* [à *Dodrans*, nine inches]. See *Dodrans*.

DOLABRIFORME *Folium* [*dolabra*, an axe] implies substance : in shape somewhat resembling an axe; *compressum, subrotundum, obtusum, extrorsum gibbum, acie acuta, inferne terctrusculum*, as in the *Mesembryanthemum dolabrif*.

DOR-

DORSALIS *Arifta* [probably for *dorfualis*, à *dorfum*, the back] Fixed to the back or external part of the *Gluma*, as in the *Avena*.

DRUPA [Δρύς *arbor*, & πίπλω, *cado*, to fall; *ripe fruit*]. A full pulpy *Pericarpium*, without *Valvulæ*, containing a ftone, as the plumb, the peach, &c. *Fruβus mollis officulo*, Tournef. *Prunus* of others.

DRUPACEÆ [*Drupa*] An order of plants in the *Fragmenta methodi naturalis* of Linnæus, containing thefe *genera*, viz. *Amygdalus, Prunus, Cerafus, Padus.*

DUMOSÆ [à *dumus*, a bufh] An order of plants in the *Fragmenta methodi naturalis* of Linnæus, containing the following *genera*, viz. *Viburnum, Tinus, Opulus, Scambucus, Rondeletia, Bellonia, Caffine, Ilex, Tomax,* &c.

DUPLICATA *Radix* [à *duplex*, double] a fpecies of the bulbofa, confifting of two folid bulbs, as in fome fpecies of the *Orchis*. Thefe roots are alfo called *Tefticulata*.
Duplicatæ Cotyledones, doubled; a fpecies of the *Dicotyledones*, exemplified in the *Malva*, and the clafs *Tetradynamia*.

DUPLICATO SERRATUM *Folium*, faw-
ed double, with leſſer teeth within the greater,
as *Water Hoarhound.* Hill.

E.

EBRACTEATUS *Racemus* [*è* priv. & *bra-*
ctea, a bracteal or floral leaf] without *bracteæ*,
as in the *Ciſtus guttatus.*

ECAUDATA *Corolla* [*è* priv. & *Cauda*, a
tail] a ſpecific term oppoſed to the elongation
of the baſe of the *Corolla* in the form of a tail,
as in the *Antirrhinum cymbalaria.*

ECHINATUM *Pericarpium* [Εχῖν⊙, *Erina-*
ceus, a hedgehog] beſet with prickles.

EFFLORESCENTIA [*effloreſco*, to blow, to
bloom] comprehends the preciſe time of the
year and month when a plant ſhews its firſt
flowers.

EMARGINATUM *Folium* [*è*, & *margo*, the
margin] deficient in its margin. When applied
to the *apex* of a leaf, it ſignifies, terminating
in a notch, the margin being diſcontinued or
broken, *quod terminatur crena.*

Emarginatum Stigma, notched, as is frequent in the clafs *Didynamia* of Linnæus.

ENERVIUM *Folium* [*è*, & *nervus*, a nerve, or ftring] having no apparent nerves. See *Nervofum*.

ENNEANDRIA [Εννεα, *novem*, nine, & ανηρ, *maritus*, a hufband] The ninth clafs in the Linnæan fyftem, comprehending fuch hermaphrodite flowers as bear nine *Stamina*. The orders are three, viz. MONOGYNIA, of which there are three *genera*, viz. *Laurus, Tinus, Caffytha*; TRIGYNIA, of which there is but one *genus*, viz. *Rheum*; HEXAGYNIA, of which there is likewife but one *genus*, viz. *Butomus*.

ENNEAPETALA *Corolla* [εννεα, novem, nine, & Πεταλλον, *petalum*] confifting of nine *petala*, as in *Thea, Magnolia, Liriodendron.*

ENODIS *Caulis, Culmus* [ex *è* & *nodus*] having no knots, or joints; *qui continuus eft, nec articulis interceptus* : oppofed to *articulatus*.

ENSATÆ [*Enfis*, a fword] An order of plants in the *Fragmenta methodi naturalis* of Linnæus, containing the following *genera*, viz. *Iris, Gladiolus,*

Gladiolus, Antholiza, Ixia, Sisyrinchium, Comme-lina, Xyris, Eriocaulon, Aphyllanthes.

ENSIFORME *Folium* [*ensis*, a sword] *est anceps a basi versus apicem adtenuatum,* shaped like a two-edged sword, gradually tapering to the point, as in the *Anthericum ossifrag. & calycula-tum.*

EQUITANTIA *Folia* [*equitans*, riding] A term in foliation, which implies the sides of the leaves being parallel, and the interior *included* by the exterior, as in the *Hemerocallis, Poa, Iris, Acorus, Carex,* & *Gramina nonnulla.*

ERECTUS *Caulis, Ramus, Folium* ; erect, upright, perpendicular ; but, when applied to a branch or leaf, must not be understood absolute-ly, but as forming an acute angle with the *Caulis,* so as to be nearly erect; *ad angulum acutis-simum cauli adsidens.*

Erecta Anthera, fixed by one extremity to the apex of the *filamentum,* opposed to *Incumbens* & *Versatilis.*

EROSUM *Folium* [ab *erodor,* to be gnawed] *cum folium sinuatum margine alios minimos obtusos acquirit;* when a sinuated leaf has its margin

broken

broken by fmaller obtufe finufes, as if gnawed or eaten.

EXSERTA *Stamina* [ab *exfero*, to put forth] when the *Stamina* appear above the *Corolla*, oppofed to *inclufa*, exemplified in the *Erica multiflora*.

EXSTIPULATUS [ab *ex*, & *ftipula*, ftubble or ftraw] without *ftipulæ*, as in the *Cardamine parviflore*.

EXSUCCUM *Folium* [*ex*, & *fuccus*, juice] regards the fubftance of leaves; dry, oppofed to *fucculentum*.

EXTRAFOLIACEÆ *Stipulæ* [*extra*, & *folium*] growing on the outfide of the leaves of the plant, as in the *Betula*, *Tilia*, *Alnus*, and the clafs *Diadelphia*.

F.

FARCTUM *Folium* [*farcio*, to ftuff, to cram] oppofed to *Tubulofum*, and *Fiftulofum*.

FASCICULATA *Folia* [*Fafciculus*, a little bundle] leaves growing in bunches or bundles,
                                                    many

many of them from the fame point, as in the *Larix, Pinus,* &c.

FASCICULARIS *Radix* [à *Faſcis,* a bundle] bundled ; a ſpecies of the Tuberoſe root, in which the knobs are collected in bundles, as in *Pæonia.*

FASCICULUS [dim. à *faſcis,* a bundle] a ſpecies of infloreſcence, in which the flowers grow erect, parallel, cloſe to each other, forming together a flat ſurface ; *colligit flores erectos, parallelos, faſtigiatos, approximatos,* as in the *Dianthus barbatus,* Sweetwilliam.

FASCIATA *Planta* [*faſcis,* a bundle] when many *caules* grow together ſo as to form a compact bundle.

FASTIGIATI *Pedunculi* [*Faſtigium,* the *apex,* or top of a pyramid, &c.] pointed at the top. Linnæus applies this term to flowers whoſe *pedunculi* are ſo proportioned in length as to form a horizontal plain, exemplified in the *Dianthi* & *Silenes.* There is a manifeſt impropriety in this application, unleſs we ſuppoſe the pyramid inverted.

FAUX, the jaws or chops ; the *hiatus* of the tube

tube of the *Corolla* ; occurring frequently in the clafs *Diàynamia* of Linnæus.

FEMINA *Planta*, female plant ; producing on the fame root female flowers only.

Femineus flos, producing *ftigmata*, but no *an-thei æ*.

FIBROSA *Radix* [à *fibra*, a fibre] a fibrous root ufed by former botanifts to fignify that kind of root which not exceeding in dimenfion the *ba-fis* of its ftem, defcends perpendicularly in one ftraight fibre, as in *Paftinaca*, *Raphanus*, &c. : but Linnæus applies it to thofe roots only which confift entirely of fmall fibres, or *Radi-culæ*.

FILAMENTOSA *Radix* [*Filum*, a thread] a fpecies of the *Fibrofa*, as in grafs.   *Hill*.

FILAMENTUM [à *Filum*, a thread] that thread-like part of the *Stamen*, which fupports the *Anthera*, and connects it with the *Receptacu-lum*.

· FILICES [à *filum*, a thread, *qu. filatim incifa*] Ferns ; one of the feven tribes or families of the vegetable kingdom, according to Linnæus, by whom it is thus characterized ; having their
fructification

fructification on the backfide of the *Frondes*. They conftitute the firft order in the clafs *Cryptogamia*, and confift of 16 *genera*, which are divided into *fructificationes spicatæ, frondosæ, &* *radicales*. This order comprehends the entire xvi[th] clafs of Tournefort, in whofe fyftem the *Filices* make only a fingle *genus*, in the firft fection of the above-mentioned clafs.

Filices, an order of plants in the *Fragmenta methodi naturalis* of Linnæus.

FILIFORMIS *Filamentum, Stylus, Receptaculum* [*Filum*, a thread, & *forma*, form or fhape] of an equal thicknefs from top to bottom, oppofed to *fubulatus*.

FIMBRICATA *Petala* [*Fimbria*, a border, or fringe] fringed, as in the *Menyanthes, Paffiflora*.

FISSUM *Folium* [à *findor*, to be cloven] cleft, or divided half-way down, its finufes being linear and margins ftraight, *finubus linearibus, marginibus rectis:* according to the number of divifions it is called *bifidum, trifidum, quadrifidum, quinquefidum, multifidum*. It differs from the *Lobatum* in its fiffures not being fo deep, nor concave, nor wide; and from *Partitum*, in being divided only half-way down.

FISTULOSUS *Caulis* [à *Fiſtula*, a pipe] a hollow ſtem, oppoſed to *Farctus*.

Fiſtuloſum folium, as in the *Oenanthe fiſtuloſa.*

FLABELLUM, a fan. Ludwig defines it to be *caulis lateralis repens vel ſub terra, vel in ejus ſuperficie, ex cujus nodis in inferiore parte fibrillæ exeunt.* Inſtitut. § 395. The *Repens caulis* of Linnæus, which ſee.

FLACCIDUS *Pedunculus*, feeble, flaccid, oppoſed to *rigidus, ſtrictus; cum ita debilis ut a proprio floris pondere dependeat.*

Flaccidus Caulis, as in the *Galium mollugo.*

FLAGELLUM [a *flagrum*, a whip or thong] a barren twig or ſhoot like a thong, as in the *Fragaria veſca ;* herbaceous as in the *Rubus ſaxatilis.*

FLEXUOSUS *Caulis, Culmus,* having many turnings; taking a different direction at every joint; *horſum verſum flexus,* as in the *Smilax.*

FLORALIA *Folia* [*Flos,* a flower] are thoſe leaves which immediately attend the flower : when they differ in ſhape or colour from the other leaves, they are termed *Bracteæ.*

FLO-

FLORALIS *Gemma* [*flos*] regards the contents of the *gemma* ; containing a flower, oppofed to *foliaris*.

FLORIFERÆ *Gemmæ* [*flos*, & *fero*, to bear] producing flowers.

FLOS, a flower. Flowers are the organs of generation of plants together with their covering. They may be either *terminales, laterales, fparfi, feffiles, pedunculati, unicus, folitarius, terni,* &c. *copiofi, erecti, cerni, nutantes, verticales,* or *horizontales.* The effential parts of a flower are the *Anthera* and *Stigma,* which conftitute its exiftence, with or without teguments.

FLOSCULUS, a little flower ; one of the diftinct flowers, or florets, which compofe a *Flos aggregatus,* an aggregate flower, as in the clafs *Syngenefia* of the fexual fyftem of Linnæus.

FOLIACEÆ *Glandulæ,* when upon the leaves : thefe are either in the ferratures, as in the *Salix* ; on the *befis,* as in the *Amygdalus, Cucurbita, Elæocarpus, Impatiens, Padus, Opulus* ; on the back, as in the *Urena, Tamarix, Croton* ; or on the fuperficies, as in the *Pinguicola, Drofera,* &c.

K

FOLIARIS *Cirrus* [*folium*, a leaf] a tendril proceeding from a leaf.

Foliaris Gemmatio, regards the contents of the *gemma*, and not its origin ; containing leaves, oppofed to *floralis*.

FOLIATIO *Plantæ* [*folium*] the complication of the leaves whilft folded within the *Gemma*, or bud : it is either *involuta, revoluta, obvoluta, convoluta, imbricata, equitantia, conduplicata, plicata, reclinata*, or *circinalia*.

FOLIATUS *Caulis* ; covered with leaves, as in the *Gladiolus*.

FOLIIFERÆ *Gemmæ* [*folium*, & *fero*, to bear] producing leaves.

FOLIOLUM [dim. of *folium*, a green leaf] one of the fingle leaves which together conftitute a *folium compofitum*.

FOLIOSUM *Capitulum* [*folium*] leafy, covered or intermixed with leaves, oppofed to *nudum*.

FOLIUM, a leaf ; the green leaf of a vegetable. Leaves, according to Linnæus, are the

lungs

lungs of plants by which they attract and tran-
fpire the air : they are *fimplex* or *compofitum*.

FOLLICULUS [dim. à *follis*, a bag] a fpecies of *Pericarpium* firft mentioned by Linnæus
in his *Delineatio Plantæ :* it is *univalvis*, or *bivalvis :* it feems to exprefs what he formerly de-
noted by *conceptaculum.*

Folliculi are little glandular veffels diftended
with air, on the furface of fome plants, as at
the root of the *Utricularia*, and on the leaves
of the *Aldrovanda :* in the firft inftance they are
vafcula bicornia, and in the latter *folliculi femicir-
culares.*

FORNICATUM *Petalum* [*Fornix*, an arch
or vault] arched, or vaulted, as in the *Lamium*,
Galeopfis, *Stachys*.

FREQUENS *Planta*, frequent, when growing fpontaneoufly in great numbers, fynonym.
with *Vulgaris*.

FRONDESCENTIA [*frons*, a leaf] comprehends the precife time of the year when a
plant firft unfolds its leaves.

FRONDOSUS *caudex* [*frons*, which fee] as
in the *Palmæ*.

Frondofus

Frondofus prolifer flos. A proliferous flower is faid to be *frondofus* when the *proles* are *foliofi,* leafy.

FRONS, a leaf or branch of a tree; ufed by Linnæus to exprefs the peculiar kind of leaves of palms and ferns ; *Trunci fpecies ex ramo coadunatus folio, & fæpius fructificatione.*

FRUCTESCENTIA [*Fructus,* fruit] comprehends the precife time of the year when a plant fcatters its ripe feeds.

FRUCTIFICATIO [*fructus,* fruit] as defined by Linnæus, *eft vegetabilium pars temporaria, generationi dicata, antiquum terminans, novum incipiens* ; the temporary part of vegetables, appropriated to generation, terminating the old vegetable and beginning the new. It confifts of feven diftinct parts, viz. *Calyx, Corolla, Stamina, Piftillum, Pericarpium, Semen, Receptaculum. Fructificatio* may be either *fimplex,* or *compofita ex flofculis.*

FRUCTIFLORÆ [*Fructus,* fruit, & *flos,* a flower] The tenth clafs in Royen's fyftem : it contains the *Coronati* of Linnæus's *Methodus Calycina.*

FRUSTRANEA *Polygamia* [*fruſtra*, to no purpoſe] The third order in the claſs *Syngeneſia* of Linnæus, containing thoſe plants in the compoſition of whoſe flowers ſome of the *floſculi* are hermaphrodite, and others neuter; in which caſe the latter are of no conſequence, the fructification being perfect in the hermaphrodites.

FRUTEX, a ſhrub. Shrubs, according to Linnæus, make a branch of the ſeventh family in the vegetable kingdom, and are diſtinguiſhed from trees in that they come up without buds; but this diſtinction is not univerſal, though it be generally juſt with regard to thoſe of Europe. Nature hath made no abſolute diſtinction between ſhrubs and trees. *Frutex*, in its general acceptation, is a plant whoſe trunk is perennial, gemmiparous, woody, dividing and ſubdividing into a great number of branches. In ſhort, it is the epitome of a tree, exemplified in the roſe-buſh.

FRUTICOSUS *Caulis* [à *Frutex*, a ſhrub] See *Frutex*.

FUGACISSIMA *Petala* [*fugax*, fleeting] of very ſhort continuance; ſoon falling off, as in the *Cardamine impatiens*.

FUL-

FULCRATUS *Caulis,* *Ramus* [*fulcio,* to prop] the branch descending to the ground, and supporting the stem, as in the *Ficus.*

FULCRUM, a prop, a support. *Fulcra,* says Linnæus, *adminiculæ sunt pro commodiore suf-tentatione :* rather, *Fulc·a* are certain minute parts of plants which serve to strengthen, support, and defend them : they are of seven different kinds, viz. *Petiolus, Stipula, Cirrhus, Pubes, Arma, Bractea, Pedunculus.*

FUNGI [à σφογγος, *fungus*] One of the seven families or tribes of the vegetable kingdom, according to Linnæus, comprehending all those which are of the mushroom kind, and which in Tournefort constitute the 2d, 3d, 4th, 5th, 6th, 7th, and 8th *genera* of the first section in the class xvii.
Fungi, an order of plants in the *Fragmenta methodi naturalis* of Linnæus.

FURCATA *frons* [à *furca,* a fork] forked, as in the *Jungermannia furcata.*
Furcata Seta, as in the *Leontodon hispidum.*

FURCÆ, forks ; a species of *Arma* growing on the surface of certain plants for their defence against external injuries : they are *bifidæ, trifi-dæ,* *dæ*, &c. according to the number of prongs of which each *furca* confifts. Thefe *Furcæ* are exemplified in the following plants, *Berberis*, *Ribes*, *Gleditfia*, *Mefembryanthemum*, *Ofteofpermum*, *Ballota*, *Barleria*, *Fagonia*, *Poterium*.

FUSIFORMIS *Radix* [à *Fufus*, a fpindle] fpindle-fhaped root, *i. e.* tapering downward to a point, as in the *Daucus*, *Paftinaca*, &c. *Fufiforme folium*, as in the *Craffula rubens*.

G.

GALEA [à γαλη, *felis*, a cat] an helmet ; a term applied to the *Corolla* of the clafs *Gynandria*, and fome of the *Dydinamia* when it is formed into the fhape of an helmet, exemplified in the *Orchis*, and in the *Pedicularis roftrata*.

GALEATUM *labium* [à *Galea*, an helmet] fhaped like an helmet, as in the *Rinanthus*.

GEMINÆ *Stipulæ*, growing in pairs, as in moft plants.

GEMINATUS *Pedunculus* [*geminus*, double] two proceeding from the fame part ; growing in pairs.

GEM-

GEMMA [*an à geno,* i. e. *gigno;* an à γεμω, *plenus fum*] a bud ; an *hybernaculum* on the afcending *caudex:* it confifts either of *Stipulæ, Petioli,* the rudiments of leaves, or cortical *fquammæ.* The various fpecies of *gemmæ* are *deciduæ, foliiferæ, foliiferæ & floriferæ diftinctæ, foliiferæ & floriferæ femineæ, foliiferæ & floriferæ mafculæ, foliiferæ & floriferæ hermaphroditæ, foliiferofloriferæ.*

GEMMATIO [*gemma,* a young bud] the formation of the *gemma* from leaves, *ftipulæ, petioli,* or *fquammæ.* With regard to its bulb, it is *folidus, tunicatus, fquammatus,* or *caulinus ;* as to its origin, it is *petiolaris, ftipularis,* or *corticalis;* and in refpect to its contents, *foliaris, floralis, communis.*

GEMMIPARUS [*Gemma,* a bud, & *pario,* to bear] producing buds.

GENERA *Plantarum,* is the fecond fubdivifion in the Linnæan fyftem : it comprehends an affemblage of fpecies, fimilar in their parts of fructification, under the fame clafs and order.

GENICULATUS *Caulis, Culmus, Pedunculus* [*genu,* the knee] in its general acceptation fignifies signifies jointed, synonymous with *articulatus*; but in Linnæus it implies the joints forming a very obtuse angle, as when the knee is a little bent; *flexuosus* in a small degree.

GENICULUM [*genu*, the knee] the little knots or joints of a *Culmus*.

GERMEN, a sprout or bud; the *basis* of the *Pistillum*; the rudiment of the fruit yet *in embryo*.

GIBBUM *Folium* [*gibba*, a hump on the back] *quod utramque superficiem facit convexam, mediante copiosiore pulpa*; when by means of the intermediate pulp both surfaces are rendered convex.

Gibbum Perianthium, regards the base of the *Perianthium*, as in the class *Diadelphia* & *Tetradynamia* of Linnæus.

GLABER, *Caulis*, *Folium*; smooth, having an even surface, *quod superficie lævi est, absque omni inæqualitate* : opposed to *Scaber*, and exemplified in the *Daphne laureola, Arbutus unedo*.

GLADIATA *siliqua* [à *gladius*, a sword or knife] shaped like a sword, as in the *Cleome arabica*.

GLANDULÆ, a species of secretory, or

excretory

excretory veffels on the furface of fome plants : they are either *petiolares, foliaceæ, ftipulares, capillares,* or *pori.*

GLANDULATIO [*glans*, an acorn, a gland], regards the fecretory veffels of plants ; thefe are *Glandulæ, Folliculi,* or *Utriculi.*

GLANDULIFERA *Scabrities,* a fpecies of fetaceous *Scabrities* on the furface of fome plants, in which there are minute glands on the extremity of each briftle, as in the *Ribes.* Lin. This is undoubtedly a very improper application of the word *Scabrities.*

GLAREOSIS, *locis* underftood [*glareo*, gravel] growing generally on a dry gravelly foil.

GLAUCOPHYLLUS [Γλαυκος, *glaucus,* blue, & Φυλλον, *folium,* a leaf] azure-leaved, as the *Canna glauca.*

GLOBOSA *Radix,* [à *globus*, a globe] a round root, as in *Bunium,* and fome fpecies of *Ranunculus,* &c.
Globofum Capitulum, a round *capitulum,* diftinguifhed from *fubrotundum, dimidiatum.*

GLOBULARIS *Scabrities* [dim. à *globus*, a round

round ball or globe] A species of glandular *Scabrities*, scarce visible to the naked eye, the small grains of which are exactly globular, on the surface of some plants, as in the *Atriplex*, *Chenopodium*, &c.

GLOCHIDES [Γλωχίς, *cuspis*, a point] the small points of the *pubes* of plants. In the *Phil. Botanica* Linnæus applies this term only to the *hami-triglochides*, three-pointed, as in the *Lappula*.

GLOMERATA *Spica* [à *glomus*, a clue of yarn or thread] indicates the flowers growing close together in somewhat of a globular form, as in the *Panicum italicum*.

Glomerata Panicula, as in the *Poa ciliaris*.

GLUMA [à *glubo*, to strip the bark from a tree] husk, chaff; a species of *calyx* peculiar to corn and grass, infolding the *arista* : it may be *uniflora*, *multiflora*, *univalvis*, *bivalvis*, *multivalvis*, *colorata*, *glabra*, *hispida*.

GLUMOSUS [from *Gluma*] applied to an aggregate flower with a filiform *Receptaculum*, whose *basis* is provided with a common *Gluma*, husk.

GLU-

GLUTINOSITAS [*gluten*, glue, paſte] a ſpecies of *Pubes*, according to Linnæus; but in what reſpect different in its ſignification from *Viſcoſitas*, I am at a loſs to determine, unleſs he intends that it ſhould mean viſcoſity in a higher degree, covered with a ſtiffer glue.

GRAMINA, graſſes; one of the ſeven tribes, or families, of the vegetable kingdom, according to Linnæus, by whom it is thus characteriſed : having the moſt ſimple leaves, an articulated *culmus*, a glumoſe *calyx*, and a ſingle ſeed. This family includes the ſeveral kinds of corn as well as graſſes. In Tournefort the *gramina* (graſſes only) make the 8th *genus* of the 3d ſection in claſs xv. *Graminum folia pecoribus & jumentis læta paſcua; Semina minora avibus, majora hominibus eſculenta ſunt.*
Gramina, an order of plants in the *Fragmenta methodi naturalis* of Linnæus.

GRANULATA *Radix* [from *granum*, a grain] granulated ; conſiſting of many little knobs attached by ſmall ſtrings, as in the *Saxifraga granulata*. Theſe roots are alſo called *aggregata.*

GYMNOSPERMIA [Γυμνος, *nudus*, naked, & Σπιρμα, ſeed] The firſt order in the claſs
Didynamia

Didynamia of Linnæus: it comprehends thofe plants, of that clafs, which have naked feeds. The feeds are conftantly four in number, except in one *genus*, viz. *Phryma*, which is *monofpcrmus.* Thefe are the *Labiati* of Tournefort, and *Ver-ticillatæ* of Ray.

GYNANDRIA [Γυνη, *mulier*, a woman, & Ανηρ, *vir*, a man] The twentieth clafs in the Linnæan fyftem; it confifts of thofe hermaphrodite plants whofe *ftamina* grow either upon the *Stylus*, or on an elongated *Receptaculum*, refembling a *Stylus*, and fupporting both *Piftillum* and *Stamina.* The firft order, viz. DIANDRIA, of this clafs, is a natural one, the *genera* differing only in the *Nectarium.* The ftructure of the parts of fructification of this order is very fingular: the *Germen* is always *contortum*; the *Petala* are five, of which the two interior generally approach fo as to form a *Galea*, whofe inferior *labium* becomes *Nectarium*, which alfo fupplies the place of a *Piftillum* and fixth *Petalum*; the *Stylus* adheres to the inferior margin of the *Nectarium*, fo that it is hardly diftinguifhable; the *Filamenta* are invariably two, fhort, fupporting two *Antheræ*, which grow narrow downward, are naked, and divifible like the pulp of the *Citrus*: thefe are included in two *cellulæ* which are open below, and adhere

to

to the interior margin of the *Nectarium*. The fruit is a *Capfula, unilocularis, trivalvis*, dividing under the carinate ribs. The feeds are *fcobiform*, numerous, fixed in each *valvula* to a linear *Receptaculum*. *Vires aphrodifiacæ omnibus his plantis ab omnibus medicis adfcribuntur*. This order has eight *genera*, viz. *Orchis, Satyrium, Ophrys, Serapias, Limodorum, Cypripedium, Epidendrum, Arethufa :* TRIANDRIA, containing but one *genus*, viz. *Sifyrinchium :* TETRANDRIA, containing but one *genus*, viz. *Nepenthes*; PENTAGYNIA, containing two *genera*, viz. *Ayena & Pafsiflora*; HEXANDRIA, containing but one *genus*, viz. *Ariftolochia*; OCTANDRIA, containing but one *genus*, viz. *Piftia*; DECANDRIA, containing but one *genus*, viz. *Helicteris*; POLYANDRIA, containing feven *genera*, viz. *Xylopia, Grewia, Pothos, Dracontium, Calla, Arum, Zoftera*.

H.

HABITUALIS *Character* [*Habitus*] The character or defcription of a plant taken from its *habitus*, which according to Linnæus confifts in the *placentatio, radicatio, ramificatio, intorfio, foliatio, ftipulatio, pubefcentia, inflorefcentia*.

HABITUS *Plantæ*, habit, external appearance, *facies externa* ; so it was understood by former botanists ; but Linnæus in the *Phil. Botan.* defines it thus, *conformitas quædam vegetabilium affinium & congenerum in placentatione, radicatione, ramificatione, interfione, gemmatione, foliatione, stipulatione, pubescentia, glandulatione, lactescentia, inflorescentia, aliisque*. In his *Delineatio Plantæ*, we find under the general title *Habitus, vernatio, æstivatio, somnus, variatio, sponsalia, feminatio*.

HAMOSA *seta* [Aμη, *falx*, a hook, asking Mr Ainsworth's pardon] hooked. Setaceous or bristly pubescence is termed *hamosa* when the *apex* of each *seta* is curved : *adhærent animalibus prætereuntibus*, says Linnæus. *Hami* are either *triglochides*, as in the *Lappula*; or *incurvi*, as in the *Arctium, Marrubium, Xanthium, Petiveria*.

HASTATUM *Folium* [*Hasta*, a spear] a leaf in shape resembling the head of an halbert, *triangulare, basi lateribusque excavatis angulis patulis*, as in the *Scutellaria hastifolia*.

HEDERIFOLIA [*Hedera*, the ivy] ivy-leaved, as in *Veronica hederifolia*.

HEMISPHERICUS *Calyx* [ex ἥμι, *semis*, half,

&

& σφαῖρα, *fphæra*, a fphere] half a fphere, as in the *Tanacetum*.

HEPTANDRIA [Επία, *feptem*, feven, & Aντε, *maritus*, a hufband] Linnæus's feventh clafs, comprehending thofe hermaphrodite flowers which have feven *Stamina* ; it has four orders, viz. MONOGYNIA, which contains two *genera*, viz. *Trientalis*, *Æfculus* ; DIGYNIA, which has but one *genus*, viz. *Limeum* ; TRIGYNIA, alfo with but one *genus*, viz. *Saururus*; HEPTAGYNIA, in which there is likewife but one *genus*, viz. *Septas*.

HERBA [*de etym. parum conflat*] an herb : according to Linnæus, it is that part of the vegetable which arifes from the root, is terminated by the fructification, and comprehends the *truncus*, *folia*, *fulcra*, & *hybernaculum*.

Herbæ, properly fpeaking, are thofe plants whofe ftems perifh annually. *Tournef*.

HERBACEÆ *Plantæ* [*herba*, an herb] are thofe plants which annually perifh down to the root ; for in the perennial kinds the *gemmæ* are produced on the root. *Lin*.

Herbaceus Caulis, indicates the time of duration of the ftem ; dying annually : not woody, oppofed to *Fruticofus*, & *Suffruticofus*.

HER-

HERMAPHRODITUS *Flos* [ab Ε*ρμης*, *Mercury*, & Αφροδιτη, *Venus*] that which contains both *Antheræ* and *Stigma* : of this kind are all Linnæus's 24 claſſes, except the 21ſt, 22d, & 24th. A plant is called *Hermaphrodita*, when on the ſame root it produces hermaphrodite flowers only. There are alſo *Flores hermaphroditi*, male hermaphrodites, and *hermaphroditæ*, female hermaphrodites : the firſt is when the *Piſtillum*, the ſecond, when the *Stamina* are abortive ; but theſe inſtances are rare.

HESPERIDÆ [*Hesperides*, whoſe orchards produced golden fruit]. An order of plants in the *Fragmenta methodi naturalis* of Linnæus, containing theſe *genera*, viz. *Citrus, Styrax, Garcinia*.

HEXAGONUS *Caulis*. See *Trigonus*.

HEXANDRIA [Εξ, *ſex*, ſix, & Ανηρ, *vir*, a man] The ſixth claſs in the Linnæan ſyſtem ; it conſiſts of thoſe plants which produce hermaphrodite flowers with ſix *Stamina*, of equal length. The orders are five, *viz.* MONOGYNIA, of which there are 51 *genera*, viz. *Bromelia, Tillandſia, Burmannia, Tradeſcantia, Pontederia, Hæmanthus, Galanthus, Leucojum, Narciſſus, Pancratium, Crinum, Amaryllis, Bulbocodium, A-*

M   *phyllanthes,*

phyllanthes, Allium, Lilium, Fritillaria, Uvula-ria, Gloriofa, Erithronium, Tulipa, Albuca, Hy-paxis, Ornithogalum, Scilla, Cyanella, Afphodelus, Anthericum, Leontice, Afparagus, Convallaria, Po-lianthes, Hyacinthus, Aletris, Yucca, Aloe, Agave, Alftroemeria, Hemerocallis, Acorus, Orontium, Ca-lamus, Juncus, Achras, Richardia, Burfera, Prinos, Berberis, Loranthus, Frankenia, Peplis; DIGYNIA, containing 3 *genera,* viz. *Velezia, Oryza, Atraphaxis;* TRIGYNIA, containing 9 *genera,* viz. *Flagellaria, Rumex, Scheuchzeria, Triglochin, Melanthium, Medeola, Trillium, Col-chicum, Helonia;* TETRAGYNIA contains but one *genus,* viz. *Petiveria;* POLIGYNIA contains but one *genus,* viz. *Alifma.* Hexandriæ radices fecundum faporem & odorem edules aut noxiæ funt: edules funt radices inodoratæ. *Lin.*

HEXAGYNIA [εξ, *fex,* fix, & Γυνη, *mulier,* a woman] One of the orders in the 9th and 13th claffes in the Linnæan fyftem; containing thofe plants in whofe fructification there are fix *Styli,* which are confidered as the female or-gans of generation.

HEXAPETALA *Corolla* [εξ, *fex,* & Πεταλον, *petalum*] confifting of fix *petala,* as in the *Tu-lipa, Lilium, Podophyllum.*

HEX-

HEXAPHYLLUS *Calyx* [Φυλλον, *folium*, a leaf] Confifting of fix leaves, as in the *Berberis*.

HIANS *Corolla*, gaping ; a fpecific diftinction, exemplified in the *Melampyrum fylvaticum*; oppofed to *claufa*.

HILUM, the black eye of a bean ; the external mark on the feed by which it was fixed to the fruit, evident in the *Cardiofpermum* & *Staphylæa*.

HIRSUTUS *calyx*, rough, hairy, as in the *Serratula alpina*.

HISPIDUS *Caulis: fetis rigidis afperfus* ; covered with ftrong fragile briftles, or prickles, but whofe roots are only fuperficial, fo as to ftrip off with the rind, as in the *Braffica erucaftrum*.
Hifpidum Folium, as in the *Turritis hirfuta*.

HOLERACEÆ [*Olus*, pot-herbs, or herbs for food] An order of plants in the *Fragmenta methodi naturalis* of Linnæus, containing thefe genera, *Spinacia, Blitum, Beta, Galenia, Atriplex, Chenopodium, Rivina, Petiveria, Herniaria, Illecebrum,* &c.

HORIZONTALIS *Flos* [*horizon*] growing with its diſk parallel to the plain of the horizon, oppoſed to *verticalis*, Or, perhaps Linnæus may mean proceeding from the ſtem in a horizontal direction.

Horizontalis Radix, a ſpecies of *Caudex deſcendens, quæ ſub terra tranſverſim extenditur*, extending horizontally, as in the *Iris*.

Horizontale folium, growing at right angles with the ſtem.

HYBERNACULUM, a place to winter in ; *eſt herbæ compendium ſuper radicem antequam excreſcens :* it is that part of the plant which incloſes and ſecures the embryo from external injuries during the winter ; it is either a *Bulbus*, or a *Gemma*.

HYBRIDA *Planta* [Υϐρις, *injuria*, injury, diſhonour] A monſtrous production of two different ſpecies, analogous to a mule in the animal creation. The ſeeds of theſe plants will not propagate. This term was uſed by former botaniſts to expreſs what Linnæus calls *Polygamia*, which ſee. *Theſium linophyllon, Antirrhinum ſpurium, Linaria*, are examples of the *hybrida*.

HYPOCRATERIFORMIS *Corolla* [υπο, *ab*,

&

& κρατηρ, *crater*, a cup] A monopetalous *Corolla* so called when the *Limbus* expands horizontally in the form of a falver, diftinguifhed, in having a *tubus*, from the *rotatus*; exemplified in the *Myofotis fcorpioides, Hottonia*.

I.

ICOSANDRIA [Εἰκοσι, *viginti*, & Ανηρ, *maritus*, a hufband] the twelfth clafs in the Linnæan fyftem, comprehending thofe plants which have hermaphrodite flowers with twenty or more *Stamina*; but the number of the *Stamina* is not to be confidered as a pofitive characteriftic. The claffical character, which diftinguifhes this from the clafs *Polyandria*, is, 1. a monophyllous concave *Calyx*; 2. the *unguis* of the *Corolla* fixed to the infide of the *Calyx*; 3. the *Stamina* above nineteen in number, and inferted in the fides either of the *Calyx* or *Corolla*. The orders are five, *viz.* MONOGYNIA, in which there are 9 *genera*, viz. *Caɛtus, Philadelphus, Pfidium, Eugenia, Myrtus, Punica, Amygdalus, Prunus, Chryfobalanus*; DIAGYNIA, containing but one *genus*, viz. *Cratægus*; TRIGYNIA, containing 2 *genera*, viz. *Sorbus, Sefuvium*; PENTAGYNIA, containing 6 *genera*, viz. *Mefpilus, Pyrus, Tetragonia, Mefembryanthemum, Aizoon,*

zoon, *Spiræa*; POLYGYNIA, containing 9 *genera*, viz. *Rosa, Rubus, Fragaria, Potentilla, Tormentilla, Geum, Dryas, Comarum, Calycanthus.* I-cosandræ fructus pulposus est esculentus. *Lin.*

IMBERBIS *Corolla*; beardless, as the *Iris*, & *Gentiana filiformis*.

IMBRICATUS *Caulis, Culmus, Calyx* [*Imbrex*, a tile] covered with, or consisting of scales in the manner of tiles upon a house. When applied to leaves, it regards their *situs*; regularly covering each other like tiles. As a term of foliation, it implies the leaves being parallel, having flat surfaces, and reciprocally covering each other, as in the *Syringa, Ligustrum, Phillyrea, Laurus, Campanula,* &c. Examples of the *Calyx imbricatus* may be seen in the *Lactuca*, and many others of the class *Syngenesia* of Linnæus.

IMMUTÆ *Cotyledones*, unaltered; a species of the *Dicotyledones*, exemplified in the *Legumina, Poma, Drupæ,* and in the class *Didynamia.*

IMPAR, odd; *cum impare*, applied to a *folium pinnatum* terminating with an odd leaf.

IN.

INÆQUALIS *Corolla*, unequal, as in the *Butomus, Salvia pratenfis*.

INANIS *Caulis* [ab *Inaniæ*, cobwebs] neither *folidus*, nor yet *fiftulofus*, but pithy.

INCANUM *Folium*, covered with a whitifh down, as in the *Draba incana*. See *Tomentofum*.

INCISUM *Folium*, cut, as in the *Ranunculus auricomus*. See *Laciniatum*.
Incifum foliolum, irregularly cut in the margin, as in the *Anemone nemorofa*, *apinnina*.

INCLINATA *Radix*, inclined ; running obliquely, as in the *Statice*. Hill.

INCLUDENS *Calyx* ; [*includo*, to include, or fhut up] fhutting up, and concealing the *Corolla*, as in the *Phalaris*.

INCLUSA *Stamina* [ab *in*, & *claudo*, to fhut in] when the *Stamina* are included within the *Corolla*, as in the *Erica vulgaris* ; it is oppofed to *exferta*.

INCOMPLETUS *Flos*, Vail. *Stamineus* of Ray. *Imperfectus* of fome botanifts ; *Apetalus* of Tournefort. See *Apetalus*.

INCRASSATUS *Pedunculus*, [*incraffo*, to make thick, to fatten] increafing in thicknefs as it approaches the flower, as in the *Cotula, Tragopogon*, and moft of the *cernuus* kind.

INCUMBENS *Anthera* [*incumbo*, to lean a-gainft] having its fide fixed to the *filamentum*, oppofed to *Erecta*.

Incumbentia Stamina, as in the clafs *Diadelphia* of Linnæus.

INCURVATUS *Caulis :* bowed. The fe-cond degree of curvature towards the earth. See *declinatus* & *nutans*.

INDIVISUM *Folium*, undivided, oppofed to *fiffum*.

INERME *Folium*, [ex *in* priv. & *arma*] un-armed, oppofed to *fpinofum, pungens*.

Inerme fpinofum, foft prickled, the edge ter-minating in foft harmlefs thorns, as *gentle thiftle*. Hill.

INFERUS *Flos* [ab *infra*, beneath] When the *receptaculum* of the flower is fituated below the *germen*, or fruit, as in the *Diandria Mo-nogynia* ; oppofed to *Superus*. It forms a divifion in fome of Tournefort's claffes under the

the title of *pistillo abeunte in fructum*, opposed to *Calyce abeunte in fructum*.

INFLATUM *Perianthium* [*in*, & *flatus*, a puff, a blast] as if blown up like a bladder; bellying out in the middle, opposed to *reflexum, tubulosum, patens*.

Inflatum Pericarpium, cum instar vesicæ cavum sit, nec repletur seminibus, as in the *Fumaria cirrhosa*.

INFLEXA *Folia* [*inflecto*, to bend inward] leaves bending upwards towards the stem, *dum sursum arcuantur versus caulem*.

INFLORESCENTIA, Inflorescence, comprehends the various modes in which flowers are joined to the plant by the *Pedunculus*; which modes are expressed by the following terms, *Verticillus, Capitulum, Spica, Corymbus, Thyrsus, Racemus, Panicula*. In the *Phil. Botanica* the terms *Axillares, Oppositifolii, Interfoliacei, Laterifolii, Petiolares, Cirrhiferi*, were also under the general head of *Inflorescentia*, but they are since transplanted under *Pedunculus loco*. Vid. *Delin. Plantæ*.

INFUNDIBULIFORMIS *Corolla* [*Infundibulum*, a funnel] monopetalous and conical,

N   with

with a tubular *basis*, as in the *Lithospermum, Cynoglossum, Pulmonaria.*

INSERTUS *Petiolus* [ab *inseror*, to be put in] as it were inserted into the stem, opposed to *adnatus, decurrens,* &c.

INTEGER *Caulis.* Linnæus, in the *Phil. Botanica,* explains this term by the word *simplicissimus,* and adds, *ramis vix ullis* ; but, in his *Delineatio Plantæ,* he gives us *simplicissimus* as a separate term ; therefore it should seem that *ramis vix ullis* belongs to *simplicissimus,* and to *integer, ramis nullis :* but then what shall we infer from these terms being, in the *Delineatio Plantæ,* ranged thus, *enodis, simplicissimus, simplex, integer ?*

Integrum folium, a leaf with an entire undivided margin, *sinu omni destitutum* ; opposed to *cordatum, lunatum, fissum, lobatum, palmatum,* &c.

Integer Calyx, as in *Genipa,* opposed to *bifidus, trifidus,* &c.

INTEGERRIMUM *Folium* [*integer,* entire] *cujus margo extimus integer absque omni crena est,* whose margin is perfectly entire, without the
leaft

leaſt notch, or *ſinus*, as in the *Rhamnus frangula*, *Trientalis europæa*.

INTERFOLIACEUS *Pedunculus* [*inter*, between, & *folium*, a leaf] proceeding from between oppoſite leaves, but ranged alternately; as in the *Aſclepias*.

INTERRUPTE *Folium pinnatum*, the ſeries of larger *foliola* being interrupted by pairs of ſmaller ones, *foliolis alternis minoribus*.
Interrupta Spica, broken, or interrupted, by intervals of leſs flowers, as in the *Mentha ſpicata*.

INTORSIO [*in*, & *torſio*, writhing] ſignifies the bending, or turning, or twiſting of any part of a plant, *flexio partium verſus alterum latus*, as *caulis*, or *cirrhus volubilis*.

INTRAFOLIACEÆ *Stipulæ* [*intra*, & *folium*] growing on the inſide of the leaves of the plant, as in the *Ficus*, & *Morus*.

INUNDATA *loca* [*in*, & *unda*, a wave, or water] according to Linnæus are places which are overflowed only in winter, *hyeme repleta aqua*, *æſtate putrida exſiccata*, *imbribus interdum ſuffuſa*.
Inundatæ, a natural order of plants which

grow

grow in the water; they form the fifth natural clafs in Scopoli's *Flora Carniolica*.

Inundatæ, an order of plants in the *Fragmenta methodi naturalis* of Linnæus.

INVOLUCELLUM [dim. ab *Involucrum*] a partial *Involucrum*; the *Calyx* of a *Pedicellus*, as in the *Euphorbia*, & *Panicum viride*.

INVOLUCRATUS *Verticillus* [*involucrum*] having an *Involucrum*.

Involucratus pedunculus, as in the *Napæa dioica*.

INVOLUCRUM [*in*, & *volvo*, to roll or wrap] that in which any thing is wrapt up; the *calyx* of umbelliferous plants, remote from the flower : it is termed *univerfale*, when below the *Umbella univerfalis*; *partiale*, when at the foot of the *Umbella partialis*; *proprium*, when belonging to a *flos umbellatus* properly fo called; and *monophyllum*, *polyphyllum*, according to its number of leaves.

INVOLUTA *Folia* [*in*, & *volvo*, to roll] when the lateral margins of the leaves, within the *gemma*, are mutually rolled fpirally inwards, as in the *Lonicera*, *Diervilla*, *Euonymus*, *Pyrus*, *Malus*, *Populus*, *Viola*, &c.

IR-

IRREGULARIS *Flos*, whofe parts want un-niformity : *Anomalus* of Tournef. and *Difformis* of others.

Irregularis Corolla, irregular, as in the *Aconi-tum*, & *Lamium*. *Irregularis*, *quæ limbi partibus, figura, magnitudine, & proportioue partium*.

JUBA, a creft, feathers : a fpecies of inflorefcence, as in the *Milium* & *Gramina*. See *Panicula*.

IULUS, a catkin, See *Amentum*.

L.

LABIATUS *Flos* [*Labium*, a lip] A mono-petalous *Corolla*, with a narrow tubular *bafis*, expanding at the top in one entire, or in two lips : *Tournef*. See *Ringens*. The *Labiati* of Tournefort are the *Verticillatæ* of Ray, and are included in the clafs *Didynamia* of Linnæus.

LACERUM *Folium* [ex λαχις, *fiffura*, a cleft or fiffure] *quod margine varie fectum eft fegmentis difformibus*, whofe margin is varioufly cut with irregular fegments ; as if rent or torn.

LACINIÆ [à *lacino*, to make holes] This term

term is applied to the *Calyx*, *Corolla*, & *Piſtil-lum*, and means the ſegments contained between the inciſions.

LACINIATUM *folium* [à *Lacinia*, a fringe or jag] *variè ſeEtum in partes, partibus iterum indeter-minate diviſis*. Divided firſt into *Laciniæ*, and thoſe again irregularly cut into ſmaller *Laciniæ*.
Laciniatus flos the ſame in Tournefort as *multifidus* in Linnæus.

LACTESCENTIA [*lac*, milk] comprehends the different coloured juices which flow copi-ouſly from particular plants on being wounded; this is either *alba*, *lutea*, or *rubra*. Lactefcentes plantæ communiter venenatæ funt, minus au-tem ſemifloſculoſæ. *Lin.*

LACUNOSUM *Folium* [*lacuna*, a ditch, a trench] deeply furrowed ; *i. e.* when the veins of the leaf are funk much below the furface; *bullatum*, in a greater degree.

LACUSTRIS *Planta* [*Lacus*, a lake] plants which grow in lakes of pure water, as the *Iſo-etes*, *Subularia*, *Plantago monanth*, *Arundo*, *Nym-phæa*, *Scirpus*, *Elatine minim.*

LAMINA, a thin plate, the tip of the ear :
the

the broad fuperior part of a polypetalous *Co-rolla.*

LANA, wool; a fpecies of pubefcens which covers the furface of many plants, ferving, according to Linnæus, as a kind of veil to fecure them from the too intenfe rays of the fun; *fervat plantas ab æftu nimio,* as in the *Salvia cana-rienfis, Sideritis canarienfis, Salvia æthiopis, Marrubium, Verbafcum, Stachys, Carduus erioce-phalus, Onopordum.*

LANATUM *Folium* [*lana,* wool] *quafi tela araneæ indutum, ut Salvia, Sideritis,* covered as with a fpider's web; fo Linnæus chufes to explain it: exemplified in the *Ledum villofum,* commonly called *cobweb ledum.*

Lanatus Caulis, as in the *Stachys germanica.*

LANCEOLATUM *Folium* [*Lanceola,* a little lance] *eft oblongum utrinque fenfim verfus extre-mitatem attenuatum;* oblong, but gradually taper-ing towards each extremity, and terminating in a point, as in the *Plantago lanceolata, Potamoge-ton lucens, crifpum,* & *ferratum.*

LATERALES *Flores* [*latus,* a fide] expreff-es a mode of inflorefcence oppofed to *termina-les;* lateral flowers.

LATERIFOLIUS *Pedunculus* [*latus*, a side, & *folium*, a leaf] proceeding from the side of the *basis* of a leaf, as in the *Claytonia, Solanum, Asperifoliæ.*

LAXUS *Caulis* ; lank, opposed to *rigidus.*

LEGUMEN, pulse ; a *Pericarpium* of two *Valvulæ*, in which the seeds are fixed along one future only, as in the *Pisum*, &c. Miller mistakes the *Legumen* for the *Siliqua.*

LENTICULARIS *Scabrities* [dim. à *lens*, a lentil] A species of glandular *Scabrities*, roughness, scarce visible to the naked eye, resembling small lentils, on the surface of some plants.

LEPROSUS [à *Lepra*, leprosy] spotted like a leper, exemplified in the *Lichen.*

LEVIS *Caulis* [rather *lævis*] smooth, having an even surface ; opposed to *striatus, sulcatus* ; as in the *Chelidonium hybridum.*
Leve Folium, as in the *Statice limonium.*

LIBER, the inner bark or rind of a tree or plant, distinct from the *cortex*, which is the outer : thus, according to Linnæus, the *calyx* is a
continuation

continuation of the *cortex*, but the *corolla* a continuation of the *liber*.

LIGNOSUS *Caulis* [*Lignum*, wood] woody,
oppofed to *herbaceus*.

LIGNUM, wood; one of the four conftituent parts of the *Radix*, according to Linnæus,
who in his *Phil. Botan.* by *radix* underftood
the ftem, together with what is generally meant
by root; but, in his *Delin. Plantæ*, he confines
the term *radix* to the root only, drops the
word *caudex* entirely, and fubftitutes *truncus* in
its ftead.

LIGULATUS *Flos* [*Ligula*, á ftráp] a fpecies of compound flower, the *corollulæ* of whofe
flofculi are tubular at the bafis, flat in the middle, and expanded towards the top. Thefe
are the *Semiflofculi* of Tournefort, and are of
the firft order in the clafs *Syngenefia*.

LILIACEÆ [*Lilium*, the lily] An order of
plants in the *Fragmenta methodi naturalis* of Linnæus, containing thefe *genera*, viz. *Lilium, Fritillaria, Tulipa, Erythronium*. The *Liliaceæ* conftitute the ninth clafs of Tournefort: they are
generally *hexapetalus*, and their *piftillum*, or *Calyx*, invariably becomes a *Capfula trilocularis*.

O   LIM-

LIMBUS, a border; the superior dilated verge of a monopetalous *Corolla*.

LINEA [*proprie est funiculus ex lino*] a line. The second degree in the Linnæan scale for measuring the parts of plants : it is the breadth of the *Lunula*, or crescent, called the *root*, on the *finger*, not the *thumb*, nail, measuring from the skin towards the body of the nail.

LINEARE *Folium* [*Linea*, a line] linear, straight, *æquali ubique latitudine*, as in the *Rofmarinus*, *Pinus*, & *Gramina*.
Linearis Caulis, when used as a term of menfuration, regards the diameter of the stem, and signifies its being extremely small. See *Linea*.

LINEATUM *Folium* [*Linea*, a line] a leaf whose superficies is slightly streaked longitudinally with parallel lines, not impressing the surface.

LINGUIFORME *Folium*, tongue-shaped. See *Lingulatum*.

LINGULATUM *Folium* [*lingua*, a tongue] or *Linguiforme* ; a leaf which is linear, carnose, obtuse, convex on the under side, and fre-
                                      quently

quently with a cartilaginous margin, as in the *Hæmanthus coccineus.*

LITHOPHYTA [Λίθος, *lapis*, a ſtone, & Φυτας, *planta*, a plant] The twentieth claſs in Royen's ſyſtem. Theſe are in fact animal productions, and therefore are improperly arranged in a vegetable ſyſtem. They are what we call in Engliſh *Corals*, &c.

LOBATUM *Folium* [λοβος, *lobus*, the lobe or tip of the ear] *diviſum ad medium in partes diſtantes, marginibus convexis* ; divided almoſt to the centre, its lobes diſtant from each other, and margins convex. A leaf is termed *bilobum, trilobum,* &c. according to the number of its lobes. The *Alchemilla vulgaris* & *minor* afford examples of the *folium lobatum.*

LOCULAMENTUM, a cell; each of the cells within that ſpecies of *pericarpium* termed *capſula,* in which the ſeeds are lodged ; *concameratio vacua pro ſeminum loco :* thus it is defined by Linnæus, but in its application it is not confined to the *capſula* only.

LOCULUS [dim. à *locus*, a place] a little cell; the minute cells which contain the *pollen* in ſome ſpecies of *antheræ.*

LOCUS *Foliorum*, place : the particular part of the plant where the leaf grows, in which re- fpect a leaf is *radicale, caulinum, rameum, axillare,* or *florale.*

LOMENTACEÆ [*Lomentum,* bean-meal] An order of plants all exotics, in the *Fragmenta methodi naturalis* of Linnæus, of which are thefe *genera, viz. Sophora, Cercis, Bauhinia, Parkinfonia, Caſſia, Poinciana, Tamarindus, Mimoſa.*

LONGIUSCULUS [dim. à comp. *longior*] rather long; a little longer than common, as in the Gramen *alopecuro accedens, petiolis longiuſculis,* Pluk. the *Agroſtis rubra* of Linnæus.

LONGUM *Perianthium,* when of an equal length with the tube of the *Corolla,* oppoſed to *Abbreviatum.*

LUCIDUM *Folium* [*Lux,* light] This is one of Linnæus's unexplained terms ; but, as it ſtands diſtinguiſhed from *nitidum,* it muſt mean clear, tranſparent.

LUNATUM *Folium* [*Luna,* the moon] moon-ſhaped, *ſubrotundum, baſi excavatum, angulis poſticis notatum.*

LUNULATA *carina* [à *lunula* dim. a half-moon] shaped like a small crescent, as in the *Polygala myrtifolia*.

LURIDÆ [*luridus*, pale, wan] An order of plants in the *Fragmenta methodi naturalis* of Linnæus, containing these *genera*, viz. *Capsicum, Solanum, Physalis, Hyoscyamus, Nicotiana, Atropa, Mandragora, Datura, Verbascum, Celsia, Digitalis:* sunt plantæ suspectæ. *Lin.*

LUTEA *Lactescentia* [à *luteum*, the yolk of an egg] yellow, as in the *Chelidonium, Bocconia, Sanguinaria, Cambogia*.
Luteus Flos, yellow, as the *Iris lutea*. Park.

LUXURIANS *Flos*, a luxuriant flower. Flowers are called *luxuriant*, when the teguments of their fructification are augmented so as to exclude some of their other essential parts. These are either *multiplicatus, plenus*, or *prolifer*. The part usually multiplied is the *Corolla*, but sometimes the *Calyx* also.

LYRATUM *Folium* [*lyra*, an harp or lyre] *est transversum divisum in lacinias, ita ut superiores majores sunt, & inferiores remotiores;* divided transversely into *laciniæ*, the superior ones being larger, and the inferior farther distant from

from each other; exemplified in the *Rumex pul-cher, Geum urbanum.*

M.

MAGNUM *Perianthium,* comparatively large, as in the *Mandragora* ; oppofed to *par-vum* & *minimum.*

MARCESCENS *Corolla* [*marceo,* to wither] withering, but not falling off, as in the *Campa-nula, Orchis, Cucumis, Cucurbita, Bryonia.*
Marcefcens Perianthium, as in the clafs *Diadel-phia* of Linnæus.

MARGO *Folii* [à *mari,* fays Ainfworth] the margin or edge of the leaf, *extrema ora folii ad latera, intaƈo difco.* A leaf, in regard to its margin, is *fpinofum, dentatum, ferratum, cre-natum, repandum, cartilagineum, ciliatum, lacerum, crofum* or *integerrimum.*

MAS *Planta* [*etym. incertum*] Male plants are thofe which on the fame root produce only male flowers, as in the clafs *Dioecia.* See *Ma-fculus.*

MASCULUS *Flos* [à *mas*] male flowers are
thofe

thofe which contain *antheræ*, but no *ftigma*. *Sterilis* of Tournefort; *Paleaceus* of Ray ; *Abortiens* of others.

MEDULLA [Μυελός, marrow] The pith or heart of the tree or plant ; according to Linnæus in his *Phil. Botan.* one of the four confti· tuent parts of the *Radix*, in which term he comprehends the ftem with what is commonly underftood to be the root : the other three parts are *lignum*, *liber*, & *cortex*, which fee.

MEMBRANACEUM *Folium* ⌊*membrana*, a membrane] a term regarding the fubftance of leaves ; having no diftinguifhable pulp between the two furfaces.
Membranacea Stipula, a thin paleaceous membrane, as in the *Arenaria rubra*.

MEMBRANATUS *Caulis*, covered with thin membranes.

MENSURA, meafure. Plants are generally fo various in their dimenfions, that their parts can only be meafured relatively to each other ; Tournefort however introduced pofitive geometrical menfuration ; but Linnæus, thinking it inconvenient for a botanift to carry an artificial fcale in his pocket, makes a natural fcale
of

of the human body, the degrees of which are thefe, *Capillus, Linea, Unguis, Pollex, Palmus, Dodrans, Spithama, Pes, Cubitus, Brachium, Orgya.*

METEORICI *Flores folares* [μετεωρος, undetermined] A fpecies of *folares* which do not obferve the hour of explication with fo much punctuality as the others, but are much influenced by fhade, the humidity of the air, weight of the atmofphere, &c.

MINIMUM *Perianthium*, comparatively very fmall, as in the *Cortufa, Dodecatheon, Patagonula, Convolvulus, Ipomoea*; oppofed to *Magnum.*

MONADELPHIA [Μόνος, *unicus*, one only, & Αδελφός, *frater*, brother] the fixteenth clafs in the Linnæan fyftem : it is a natural clafs, and comprehends thofe plants which produce hermaphrodite flowers with one collection of united *Stamina.* The natural characteriftics are CALYX, *Perianthium*, always prefent, continuing, and frequently double. The *Calyx* merits peculiar attention, being, in this clafs, the pofitive generical diftinction. COROLLA, *Petala* five, obcordate, the fides lapping each over the other, contrary to the motion of the fun. STAMINA. *Filamenta* united below, diftinct

tinct above, the exterior ones fhorter : *Antheræ* incumbent. PISTILLA, *Receptaculnm* of the fructification, prominent in the centre of the flower ; *Germina*, erect, rotato-articulate, furrounding the *apex* of the *receptaculum* ; *Styli*, all united below in one body with the *Receptaculum*, and at the top diftinguifhed into as many *filamenta* as there are *germina* ; *Stigmata*, fpreading, and flender. PERICARPIUM, *Capfula* divided into as many diftinct *locula*, as there are *piftilla*. SEMINA, reniform. The plants of this clafs are emollient, and mucilaginous. The orders are three, *viz.* PENTANDRIA, containing 4 *genera*, viz. *Waltheria, Hermannia, Melochia, Bombax* ; DECANDRIA, containing 3 *genera*, viz. *Connarus, Hugonia, Geranium* ; POLYANDRIA, containing 14 *genera*, viz. *Adanfonia, Sida, Napæa, Althæa, Alcea, Malva, Lavatera, Malope, Urena, Goffypium, Hibifcus, Pentapetes, Stewartia, Camellia.*

MONANDRIA [µόνος, *unicus*, one, & ανηρ, *maritus*, a hufband] The firft of Linnæus's twenty-four claffes : it comprehends thofe plants which produce hermaphrodite flowers, having but one *ftamen* ; its orders are two, *viz.* MONOGYNIA, which comprehends 11 *genera*, viz. *Canna, Amomum, Coftus, Alpinia, Maranta, Curcuma, Kæmpferia, Thalia, Boerhavia, Salicornia,*

nia, Hippuris ; DIGYNIA, in which are four *genera*, viz. *Corifpermum, Callitriche, Blitum, Cinna.*

MONANGIÆ [μονος, *unicus*, & Αγϝ۞, *vas*, a veſſel, or *loculamentum*] The fifteenth claſs in Boerhaave's ſyſtem ; it contains two orders, viz. *Primula, & Lichynis.*

MONOCOTYLEDONES [μονος, *unicus*, one, & *cotyl.*] A term of placentation, applied to thoſe plants whoſe feeds have but a ſingle *Cotyledon*, which remains within the feed : theſe are either *perforatæ*, as in *Gramina*; *unilaterales*, as in *Palmæ* ; or *reduсtæ*, as in the *Cepa.*

MONOECIA [μονος, *unicus*, one, & Οῖκ۞, *domus*, a houſe] The twenty-firſt claſs in the Linnæan ſyſtem, comprehending the *androgynus* plants, *i. e.* thoſe which produce both male and female flowers, having no hermaphrodite ones. The orders are eleven, *viz.* MONANDRIA, containing 4 *genera*, viz. *Zannichellia, Ceratocarpus, Hippomane, Cynomorium* ; DIANDRIA, containing but one *genus*, viz. *Lemna* ; TRIANDRIA, containing 12 *genera*, viz. *Zea, Tripfacum, Coix, Olyra, Carex, Sparganium, Typha, Axyris, Phyllanthus, Tragia, Hernandia, Omphalea* ; TETRANDRIA, containing 4 *genera*, viz. *Urtica, Morus,*

Morus, Buxus, Beluta ; PENTANDRIA, con-
taining 5 *genera*, viz. *Xanthium, Ambrofia, Par-
thenium, Iva, Amarantus* ; HEXANDRIA, con-
taining 3 *genera*, viz. *Zizania, Pharus, Solandra* ;
HEPTANDRIA, containing but one *genus*, viz.
Guettarda ; POLYANDRIA, containing 12 *genera*,
viz. *Sagittaria, Myriophyllum, Ceratophyllum, The-
ligonum, Poterium, Fagus, Quercus, Juglans,
Corylus, Carpinus, Platanus, Liquidamber* ; Mo-
NADELPHIA, containing 10 *genera*, viz. *Hura,
Pinus, Cupreffus, Thuja, Acalypha, Plukenetia,
Croton, Ricinus, Jatropha, Sterculia* ; SYNGENE-
SIA, containing 6 *genera*, viz. *Trichofanthes, Mo-
mordica, Cucumis, Cucurbita, Sicyos, Bryonia* ;
GYNANDRIA, containing but one *genus*, viz.
Andrachne.

MONOGYNIA [μονος, *unicus*, & Γυνη, *mulier*, a woman] The firft order in each of the
firft thirteen claffes in the Linnæan fyftem :
it diftinguifhes thofe plants in whofe fructification there is but one *Piftillum*, which is confidered as the female organ of generation.
Where the *Stylus* is wanting, it regards the
Stigma.

MONOPETALA *Corolla*, [μονος, & Πεταλον,
petalum] confifting of but one *petalum*, as the
Convolvulus, Primula.

MO-

MONOPHYLLUM *Involucrum* [μονος, *uni-cus*, one, & Φυλλον, *folium*, a leaf] confifting of a fingle leaf.

Monophyllus Calyx, as in *Datura, Primula.*

MONOSPERMA [μονος, & *fperma*, feed] having a fingle feed, as the *Polygonum, Collinfonia.*

MILIARIS *Scabrities* [*milium*, a fmall grain called *millet*] a fpecies of glandular *Scabrities*, on the furface of fome plants, like grains of millet.

MUCRONATUM *Folium* [à μαχρος, *longus*, long] *Mucro* fignifies the fharp point of a fword or dagger; fo that *mucronatum*, regarding the the *apex* of a leaf, indicates its terminating in a point, as in the *Bromelia ananas.*

Mucronatus Scirpus, the pointed bulrufh.

MULTIFIDUM *Folium* [ex *multus*, many, & *findo*, to cleave, or divide] divided into many parts, by linear *finufes* and ftraight margins. See *Fiffum.*

Multifidus Flos : Laciniatus of Tournefort ; *Monopetaloides* of others.

MULTIFLORUS *Pedunculus* [*multus*, many, &

& *flos*, a flower] bearing many flowers : producing many fructifications on each *Pedunculus*.

Multiflorus Calyx, common to many *flosculi*, as in *Scabiosa*, & the clafs *Syngenesia*.

MULTIPARTITUM *Folium* [*multus*, many, & *partitus*, divided] confifting of many divifions, *usque ad basin*, down to the bafe.

MULTIPLICATÚS *Flos*, a luxuriant flower, whofe *Corolla* is multiplied fo as to exclude fome of the *Stamina*. When the *Stamina* are, by the multiplication of the *Corolla*, entirely wanting, it is then called *Plenus*, and not *Multiplicatus*. *Flores multiplicati*, are either *duplicati*, *triplicati*, or *quadruplicati*, according to the number of repetitions of the *Corolla*. Monopetalous flowers are frequently found multiplied, but rarely full, *pleni*. Take care, fays Linnæus, left you miftake a coloured *Perianthium* for a multiplication of the *Corolla*. A multiplicate flower may always be diftinguifhed from a compound one, by its having only one *Pistillum* common to the whole.

MULTISILIQUÆ [*multus*, many, & *Siliqua*, a pod] An order of plants in the *Fragmen*

ta

ta methodi naturalis of Linnæus, amongſt which are the following *genera*, viz. *Pæonia, Aquilegia, Aconitum, Nigella, Helleborus,* `Ranunculus, Anemone,* &c.

MURICATUS *Caulis* [*murex*, a fiſh whoſe ſhell is covered with ſharp points, or prickles] prickly.

Muricatæ, an order of plants in the *Fragmenta methodi naturalis* of Linnæus, in which there are the following *genera*, viz. *Bromelia, Renealmia, Tillandſia, Burmannia.*

Muricata Semina, as in the *Caucalis.*

Muricatus Calyx, as in the *Crepis biennis.*

MUSCI [Μοσχος, *vitulus* ; properly any thing young, new, or freſh] moſſes ; one of the ſeven tribes or families of the vegetable kingdom, according to Linnæus, and by him thus characterized, having *anthera* without *filamenta*, remote from the female flower; no *piſtillum* ; and ſeeds without either *arillus* or *cotyledon*. They conſtitute the ſecond order in the. claſs *Cryptogamia*, and comprehend eleven ſpecies, which are divided into *acalyptrati*, *calyptrati diclini*, and *calyptrati monoclini*. In Tournefort the moſſes conſtitute the firſt *genus* of the firſt ſection of claſs xvii.

Muſci,

Mufci, an order of plants in the *Fragmenta methodi naturalis* of Linnæus.

MUTICA *gluma* [à *mutilus*, broken off] when the *arifta* is wanting, oppofed to *Ariftatus*; exemplified in feveral fpecies of the *Agroftis*, viz. *ftolonifera, paluftris, capillaris, fylvatica, minima.*
   Muticus Calyx, as in the *Serratula.*
   Mutica Panicula, as in the *Aira arundinacea.*
   Mutici Flofculi, as in the *Aira indica.*
   Mutici Flores, as in the *Aira criftata.*

MUTILATUS *Flos*, a mutilated flower, according to Linnæus, is a flower wanting its *Corolla*, which is generally owing to its want of proper heat.

N.

NATANS *Folium* [*nato*, to fwim] applied to aquatic plants; fwimming on the furface of the water, as in the *Nymphæa, Potomogeton.*

NAVICULARIS *Valvula* [dim. à *navis*, a fhip] in fhape refembling a little fkiff, as in the *Ifatis, Thlafpi.*

NECESSARIA *Polygamia*, neceffary. The fourth

fourth order in the clafs *Syngenefia* of Linnæus, comprehending thofe plants in the compofition of whofe flowers fome of the *flofculi* are male and the others female.

NECTARIUM [from *nectar*, honey] according to Linnæus, the melliferous part of the *corolla*, proper to the flower, and by him firft diftinguifhed from the petals ; but though it generally makes part of the *corolla*, yet it is often diftinct from it ; when this is the cafe, it is found remarkably various, and in general the plants are poifonous, as for inftance, in *Aconito*, *Helleboro*, *Aquilegia*, *Nigella*, *Parnaffia*, *Epimedio*, *Clutia*, *Meliantho*, &c.

NERVOSUM *Folium* [*Nervus*, a nerve or ftring] when their veffels neither branch nor anaftomofe, but extend in fimple lines or curves from the *bafis* to the *apex*; *quum vafa fimpliciffima abfque ramulis extenduntur a bafi verfus apicem.*

NIDULANTIA *femina*, *Bacca* [*nidus*, a neft] *per pulpam fparfa*, difperfed among the pulp, as in the *Nymphæa*.

NITIDUM *Folium*; bright, fhining, gloffy ; *quod glabritie lucidum eft*, as in the *Ferula canadenfis*, *Angelica canadenfis*.

NU-

NUCAMENTACEÆ [*Nucamentum*, a cat's tail, or long excrefcence hanging down from the pine, fir, *&c.*] An order of plants in the *Fragmenta methodi naturalis* of Linnæus, containing the following *genera*, viz. *Xanthium*, *Ambrofia*, *Parthenium*, *Iva*, *Micropus*, *Artemifia*.

NUCAMENTUM. See *Amentum*.

NUCLEUS, a kernel.

NUDUS *Caulis* : naked. Linnæus, in his *Philofophia Botanica*, explains this word by *foliis deftitutus* ; but as, in his *Delineatio Plantæ*, he makes it a diftinct term from *Aphyllus*, he muft intend that it fhould imply entire nakednefs, in oppofition not to *foliatus* only, but alfo to *fcaber*, *villofus*, &c. When applied to leaves, it is evidently taken in this fenfe.

Nudus Verticillus, having no *Involucrum*, oppofed to *involucratus*.

Nudum Capitulum, having no leaves, oppofed to *foliofum*.

NUTANS *Caulis*, nodding ; the third, and greateft degree of curvature towards the earth. See *declinatus*, *incurvatus*.

Nutans flos, when the *pedunculus* is confiderably curved.

Q   NUX,

NUX, a nut; a feed covered by a fhell, *teƈtum epidermide offca.* Lin.

O.

OB, in compofition, for *obverfus,* turned, *e. g. obcordatum folium,* the heart-fhaped leaf, fixed by its *apex,* inftead of its bafe, to the *petiolus :* and fo of *obovatum,* &c.

OBCORDATUM *Petalum* [*ob,* & *cordatum,* heart-fhaped] *Cordatum* with its *apex* downwards, as in the clafs *Monadelphia* of Linnæus.
Obcordata Siliqua, as in the *Thlafpi.*
Obcordatum Legumen, as in the *Polygala.*

OBLIQUUM *Folium* [ex *ob,* & *liquus,* tranf-verfe] when the *apex* of the leaf points to the horizon, and the *bafis* upwards, as in the *Protea, Fritillaria.* When applied to *Caulis,* it means having an oblique direƈtion, but not curved.

OBLONGUM *Folium;* a leaf whofe longitudinal diameter is much longer than its tranf-verfe, both extremities being rounded, but narrower than the fegment of a circle, as in the *Rumex acetofa,* & *Ceraftium tomentofum.*

OB-

OBSOLETE *lobatum folium*, inelegantly lobated, or divided into lobes fcarce difcernible, as in the *Jungermannia trilobata*, and the *Malva rotundifolia*.

OBTUSUM *Folium* [*obtundor*, to be blunted at the point] having a rounded *apex*, *quod terminatur quafi intra fegmentum circuli*, oppofed to *acutum*.

Obtufum Perianthium, as in the *Convolvulus, Melia*.

Obtufa Capfula, as in the *Rhinanthus*.

OBVOLUTUM *Folium* [*ob*, & *volvo*, to roll] A term in foliation, fignifying that the margins of the leaves alternately embrace the ftraight margin of the oppofite leaf, *quorum margines alterni comprehendunt oppofiti folii marginem rectum*; as in the *Dianthus, Lychnis, Saponaria, Valeriana, Marrubium, Salvia*, &c.

Obvolutæ Cotyledones, rolled up; a fpecies of the *Dicotyledones*, exemplified in the *Helxine*.

OCTANDRIA [Ὀκτω, *octo*, eight, & ανηρ, *maritus*, a hufband] 1 he eighth clafs in the Linnæan fyftem, comprehending hermaphrodite flowers with eight *Stamina*. in this clafs there are four orders, viz. MONOGYNIA,

which

which includes 26 *genera*, viz. *Tropæolum, Os-beckia, Rhexia, Oenothera, Gaura, Epilobium, Melicocca, Amyris, Griflea, Combretum, Alophyl-lus, Ximenia, Mimusops, Jambolifera, Santalum, Memecylon, Lawsonia, Vaccinium, Erica, Daphne, Dirca, Gnidia, Stellera, Pafferina, Lachnea, Bæckea*; DIGYNIA, containing 3 *genera*, viz. *Galenia, Weinmannia, Moehringia*; TRIGYNIA, containing 5 *genera*, viz. *Polygonum, Coccoloba, Paullinia, Cardiofpermum, Sapindus*; TETRA-GYNIA, containing 3 *genera*, viz. *Paris, Adoxa, Elatine.*

OCULUS, an eye: the *gemma* fo called, by Ludwig, when proceeding from the *alæ* of leaves.

OFFICINALIS [*Officina*, a fhop] of the fhops; ufed in medicine, and therefore kept in the fhops of apothecaries, as *Valeriana offi-cinalis*, &c. Ufed to diftinguifh the fpecies of particular plants.

OLIGANTHERÆ [Ολίγ⊙, *exiguus*, fmall, few, & *Anthera*] The fixteenth clafs in Royen's fyftem: it contains thofe plants whofe *petala* or fegments equal or exceed their number of *Stamina*; hence it includes many of the plants in Linnæus's firft five claffes.

OPER-

OP

OPERCULATA *Anthera*, [*Operculum*, a co-
ver] See *Operculum*.

OPERCULUM, a cover, lid; a fpecies of
cover to the *antheræ* of the *Mufci*, as in the
Sphagnum, &c.

OPPOSITI *Rami, Folia*; branches or leaves
growing oppofite to each other in pairs; when
applied particularly to branches, it implies each
pair croffing that above and below it; oppofed
to *Alterni*. Among many other plants, the
Bartfia alpina affords an example of *folia oppofita*.

OPPOSITIFOLIUS *Pedunculus* [*oppofitum,*
oppofite, & *folium*, a leaf] growing oppofite to
the leaf, as in the *Piper, Saururus, Phytolacca,
Dulcamara, Vitis, Ranunculus aquatilis, Geranium,* &c.

OPPOSITIVÆ *petiolares Gemmæ*, for *oppo-
fitæ*; oppofite, as in the *Liguftrum, Phillyrea,
Nyctanthes, Syringa, Hypericum, Buxus, Jafminum, Laurus,* &c.
Oppofitivæ ftipulaceæ Gemmæ, oppofite, as in
the *Cephalanthus, Rhamnus catharticus.*

ORBICULATUM *Folium*, [*Orbis*, an orb,
or

or circle] *cujus diameter longitudinalis & transver-salis æquales, peripheria circinata :* in plain Eng-lish, a round leaf, regarding the circumscription only, exemplified in the *Rumex digynus.*

ORCHIDEÆ [*Orchis*, the first *genus* in the class *Gynandria*] An order of plants in the *Fragmenta methodi naturalis* of Linnæus : it consists of the following *genera*, viz. *Orchis, Satyrum, Serapias, Herminium, Neottia, Ophrys, Cypripedium, Epidendrum, Limodorum, Arethusa.* Orchideæ sunt aphrodisiacæ. *Lin.*

ORDO, order ; the first subdivision in the Linnæan system of plants. In the first thirteen classes it is determined by the number of the *Pistilla* or female parts of generation, and signified by the Greek word Γυνή, *mulier*, a woman, compounded with the numerical terms, μόνος, δὶς, &c. as for instance, *Monogynia*, one woman, *Digynia*, two women, &c. The number of the *Pistilla* is generally taken from the *basis* of the *Stylus* ; but where the *Stylus* is deficient, we must estimate by the *Stigmata.* The orders in the remaining classes are determined by distinctions in the fruit, the *Pericarpium*, the *Stamina*, complication of sexes, *&c.*

ORGYA, Οργυία, *idem*] The last degree in the

the Linnæan scale for measuring plants : the distance between the extremities of the two middle fingers when the arms are extended ; or six Parisian feet. See *Mensura.*

ORGYIALIS *Caulis* [from Ὀργυιὰ, *orgyia*, 6 foot]. See *Orgya.*

OVALE *Folium,* oval ; *cujus diameter longitudinalis superat transversalem, superiore, & inferiore extremitate angustiore :* elliptical, regarding the circumscription only.

OVARIUM [ab *ovum*] The *germen* of Linnæus so called by Ludwig, &c. See *Germen.*

OVATUM *Folium* [ab *ovum,* an egg] *cujus diameter longitudinalis superat transversalem, basi segmento circuli circumscripta, apice vero eodem angustiore :* a leaf resembling in circumscription the longitudinal section of an egg, exemplified in the *Epilobium montanum, Vaccinium myrtillus, Arenaria peploides, & trinervia.*

P.

PAGINA *Folii* [*Pagina,* the page of a book] the

the fuperficies of a leaf, fuperior and inferior, *fupinum, vel pronum.*

PALEA, chaff; a thin membrane, fpringing from a *Receptaculum Commune,* which feparates the *Flofculi* from each other.

PALEACEUS *Pappus* [*palea,* fhort ftraw, or chaff] chaffy, as in the *Bidens, Silphium, To-getes, Coreopfis,* &c.

Paleaceus Flos, Ray. *Mafculus* of Linnæus.

Paleacea Gluma, as in the *Schoenus, Cyperus, Scirpus, Eriophorum.*

Paleaceum Receptaculum, as in the *Dipfacus,* & *Scabiofa.*

PALMÆ [à Παλάμη, the palm of the hand] palms; one of the feven families, or tribes of the vegetable kingdom, according to Linnæus, by whom they are thus characterized, *caudex fimplex, apice frondofus, fructificationes in fpadice cum fpatha.* We find them in the appendix to his *Syftema Naturæ,* confifting of nine *genera,* which are divided into *flabellifoliæ, pennatifoliæ,* and *bipennatifoliæ.*

PALMARIS *Caulis* [à *Palmus*]. See *Palmus.*

PALMATA *Radix* [à *Palma,* a hand] a root which

which in the fpreading of its fibres refembles a hand, as in the *Orchis*. Linnæus makes it a fpecies of the Tuberofe, I think, with very little propriety.

Palmatum Folium, refembling an open palm or hand, *longitudinaliter in partes plures fubæquales divifum verfus bafin, qua tamen cohærent in unum*; a fimple leaf with feveral divifions, almoft down to its *bafis*, as in the *Geranium pufillum*, *Ranunculus fceleratus*.

PALMUS [*Palma*, the palm of the hand] The fifth degree in the Linnæan fcale for meafuring the parts of plants : the breadth of the palm meafuring from the thumb, or three Parifian inches. See *Menfura*.

PALUSTRIS [*Palus*, a fen or marfh] growing in marfhy or fenny places : the trivial name of a fpecies of the *Scirpus* and many others.

PANDURIFORME *Folium* [*Pandura*, a mufical inftrument] fhaped like a Spanifh guitar, *oblongum inferne latius, lateribufque coarctatum*.

PANICULA [à *panus*, a woof about the quill in the fhuttle] a fpecies of inflorefcence

R   refembling

refembling a head of grain on the *Avena*, the outline of the entire figure being nearly elliptical; *fructificatio fparfa in pedunculis diverfe fubdivifis*, a fructification difperfed on *pedunculi* varioufly fubdivided : it is either *diffufa* or *coarctata*.

PAPILIONACEUS [à *Papilio*, a butterfly] butterfly-fhaped. A flower is termed *papilionaceous* when confifting of four *petala*, the upper one fpreads, the under one refembles a boat, and the two fide *petala* ftand fingle : fuch are the entire clafs *Diadelphia* of Linnæus, the *Papilionacei* of Tournefort, the *Irregulares tetrapetali* of Rivinus, and the *Leguminofæ* of Ray. The different *Petala* are termed *Vexillum, Alæ*, & *Carina*.

Papilionaceæ, an order of plants in the *Fragmenta methodi naturalis* of Linnæus, amongft which are thefe *genera*, viz. *Erythrina, Anagyris, Robinia, Lupinus, Clitoria, Vicia, Tragacantha, Lotus*, &c.

PAPILLOSUM *Folium* [*papilla*, the nipple] a leaf whofe furface is covered with dots or points like little bladders, *quod tegitur punctis veficularibus*.

PAP-

PAPPUS, down ; a fpecies of *coronula feminis*, which is either *capillaris, plumofus, paleaceus*, or *ftipatus*.

PAPULOSUM' *Folium* [*papula*, a pimple] a leaf whofe furface is covered with pimples. This fhould feem to be the natural meaning of the word when applied to a leaf : otherwife, *papula*, as ufed by Celfus, fignifies a larger fwelling covered with fmall pimples.

PARABOLICUM *Folium* ; *cujus diameter longitudinalis fuperat tranfverfalem, & a bafi furfum anguftatur in femiovatum* ; whofe longitudinal diameter exceeds the tranfverfe, and growing narrower from the bafe upwards becomes *femiovatum*. See *Ovatum*.

PARALLELUM *Diffepimentum*, parallel to the fides of the *Pericarpium*, oppofed to *tranfverfum* ; exemplified in the *Draba*.

PARASITICUS *Caulis* [*parafitus*, a parafite] *alteri plantæ, nec terræ innatus*, growing upon another plant, as the *Epidendron, Vifcum, Tillandfia*, &c.

PARTIALIS *Umbella*, a partial umbel ; the *umbellula*, or little *umbella*, which ftands upon

the

the *apex* of each *pedunculus* of an *umbella compo-sita*.

Partiale Involucrum, when at the foot of the *Umbellula*.

PARTITUM *Folium*, a divided leaf ; *divi-sum usque ad basin*, divided down to the base. Leaves are termed *bipartitum*, *tripartitum*, &c. according to the number of divisions.

PARVUM *Perianthium*, comparatively little, as in the *Theophrasta*, *Spigelia*, *Azalea*, *Lonicera*, &c. opposed to *Magnum*.

PATENS *Caulis*, *Ramus*, &c. ; spreading. Applied to *folium*, *quod ad angulum acutum cauli insidet*, says Linnæus, in his *Phil. Botanica*, placing this term between *Erectum* and *Horizontale* ; forming an acute angle with the stem, *i. e.* an angle something less than 45 degrees.

Patens Calyx, as in the *Siriapis*.

PATULUS *Calyx*, spreading, as in the *Ranunculus repens*, *acris*.

PAUCIFLORIS [*pauci*, few, & *flos*, a flower] having few flowers, as the *Veronica montana*.

PEDALIS *Caulis* [à *Pes*, a foot].   See *Pes*.
PE-

PEDATUM *Folium* [*pes*, a foot] or *Ramo-sum*, branching ; *cum petiolus bifidus latere tantum interiore adnectit foliola :* a fpecies of *folium compofitum*, in which a bifid *petiolus* connects *foliola* on its interior fide only, fomewhat refembling a bird's foot, as in the *Paffiflora, Arum, Helleborus fœtidus.*

PEDICELLUS [à *pediculus*, a little foot] *eft Pedunculus partialis*, i. e. the little foot-ftalk which fupports each feparate flower when there are more than one fructification on one *Pedunculus.*

PEDICULUS [dim. à *Pes*, a foot] The foot-ftalk of a flower, or leaf, according to former botanifts ; the *Pedunculus* of Linnæus.

PEDUNCULARIS *Cirrus* [*pedunculus*] a tendril proceeding from the foot-ftalk of a flower.

PEDUNCULATI *Flores* [*pedunculus*] growing upon foot-ftalks, oppofed to *Seffiles.*

PEDUNCULUS [from *pedo*, one who is fplay-footed] the foot-ftalk of a flower, diftinguifhed from that of a leaf ; *truncus partialis, elevans fructificationem, nec folia.* A *Pedunculus* may be either *pedicellus*, or *communis* ; with regard

gard to the place of its infertion, *radicalis, caulinus, axillaris,* &c. ; in refpect to difpofition, *alternus, fparfus, oppofitus,* &c. ; as to number, *geminatus, umbellula,* &c. ; direction, *adpreffus, erectus,* &c.; *uniflorus, biflorus,* &c. according to the number of fructifications upon each *Pedunculus* ; as to ftructure, *teres, triqueter, filiformis, articulatus,* &c.

PELTATUM *Folium* [*Pelta,* a target] the *Petiolus* being inferted into the difk of the leaf, and not into its bafe or margin, as in the *Nymphæa, Hernandia, Colocafia, Hydrocotyle vulgaris.*

PENDULA *Radix* [à *pendeo,* to hang] a pendulous root; a fpecies of the Tuberofa, in which, when the root is lifted up, the knobs hang down, as in the *Filipendula,* and *Elæagnus.*
Pendulus Pedunculus, hanging down, oppofed to *erectus, ftrictus.*
Pendula Corolla, as in the *Lathræa fquammaria.*

PENICILLIFORMIA *Stigmata* [à *penicillus,* a pencil] in fhape refembling a painter's pencil, as in the *Milium.*

PENTAGONUS *Caulis.* See *Trigonus.*
Pentagonus Calyx, as in the *Chenopodium.*

PEN-

PENTAGYNIA [πεῖε, *quinque,* five, & Γυνη, *mulier,* a woman] one of the orders in the 5th, 10th, 11th, 12th, 13th, and 20th claſſes in the Linnæan ſyſtem : in thoſe claſſes it diſtinguiſhes the plants in whoſe fructification there are five *Piſtilla,* which are conſidered, in the ſexual ſyſtem, as the female organs of generation.

PENTANDRIA [πεῖε, *quinque,* five, & ανηε, *maritus,* a huſband] the fifth claſs in the Linnæan ſyſtem ; it comprehends ſuch flowers as have five *Stamina* : its orders are ſix, *viz.* Mo· ṄOGYNIA, which contains 122 *genera,* viz. *Heliotropium, Myoſotis, Lithoſpermum, Anchuſa, Cynogloſſum, Pulmonaria, Symphytum, Cerinthe, Onoſma, Borago, Aſperugo, Lycopſis, Echium, Tournefortia, Nolana, Diapenſia, Aretia, Androſace, Primula, Cortuſa, Soldanella, Dodecatheon, Cyclamen, Menyanthes, Hottonia, Hydrophyllum, Lyſimachia, Anagallis, Theophraſta, Patagonula, Spigelia, Ophiorrhiza, Randia, Azalea, Plumbago, Phlox, Convolvulus, Ipomoea, Polemonium, Campanula, Roella, Phyteuma, Trachelium, Samolus, Nauclea, Rondeletia, Macrocnemum, Bellonia, Portlandia, Cinchona, Pſychotria, Coffea, Chiococca, Hamellia, Lonicera, Trioſteum, Morinda, Conocarpus, Erithalis, Menais, Genipa, Muſſænda, Mirabilis, Coris, Verbaſcum, Datura, Hioſcyamus, Nicotiana, Atropa, Phyſalis, Solanum, Capſicum, Strychnos, Jaquinia,*

Jaquinia, Chironia, Cordia, Ehretia, Varronia, Laugieria, Brunsfelfia, Ceftrum, Lycium, Chryfophyllum, Sideroxylon, Rhamnus, Phylica, Ceanothus, Bittneria, Myrfine, Celaftrus, Evonymus, Diofma, Hartogia, Brunia, Itea, Galax, Cedrela, Mangifera, Cupania, Hirtella, Ribes, Gronovia, Hedera, Vitis, Lagoecia, Sauvagefia, Claytonia, Achyranthes, Celofia, Illecebrum, Glaux, Thefium, Rauvolfia, Cerbera, Vinca, Gardenia, Nerium, Plumeria, Echites, Cameraria, Tabernæmontana, Ceropegia ; DIGYNIA, containing 66 *genera,* viz. *Periploca, Cynanchum, Apocynum, Afclepias, Stapelia, Herniaria, Chenopodium, Beta, Salfola, Anabafis, Creffa, Trianthema, Gomphrena, Bofea, Ulmus, Nama, Hydrolea, Heuchera, Swertia, Gentiana, Phyllis, Eryngium, Hydrocotyle, Sanicula, Aftrantia, Bupleurum, Echinophora, Tordylium, Caucalis, Artedia, Daucus, Ammi, Bunium, Conium, Selinum, Athamanta, Peucedanum, Crithmum, Cachrys, Ferula, Laferpitium, Heracleum, Ligufticum, Angelica, Sium, Sifon, Bubon, Cuminum, Oenanthe, Phellandrium, Cicuta, Æthufa, Coriandrum, Scandix, Chærophyllum, Imperatoria, Sefeli, Thapfia, Paftinaca, Smyrnium, Anethum, Carum, Pimpinella, Apium, Ægopodium, Haffelquiftia ;* TRIGYNIA, containing 15 *genera,* viz. *Rhus, Viburnum, Caffine, Sambucus, Spathelia, Staphylea, Tamarix, Turnera, Telephium, Corrigiola, Pharnaceum, Alfine, cine*, *Drypis*, *Basella*, *Sarothra* ; TETRAGYNIA, containing 2 *genera*, viz. *Parnassia*, *Evolvulus*; PENTAGYNIA, containing 8 *genera*, viz. *Aralia*, *Barrera*, *Statice*, *Linum*, *Aldrovanda*, *Drosera*, *Crassula*, *Sibbaldia* ; POLYGYNIA, containing but one *genus*, viz. *Myosurus*.——The word *Pentandria* is also an ordinal title in Linnæus's 16th, 18th, 20th, 21st, and 22d classes.

PENTANGIÆ [πεντε, five, & αγγος, *vas*, a vessel, or *loculamentum*] The nineteenth class in Boerhaave's system.

PENTAPETALA *Corolla* [πεντε, *quinque*, & πεταλον, *petalum*] consisting of five *petala*, as in the *Umbellatæ*, viz. *Pentandria*, *Digynia*.

PENTAPHYLLUS *Calyx* [πεντε, *quinque*, & φυλλον, *folium*, a leaf] consisting of five leaves, as in the *Cistus*, *Adonis*, *Cerbera*.

PERENNIS *Radix* [ex *per*, by, & *annus*, a year] a perennial, or continual root ; *i. e.* continuing several years.

PERFECTUS *Flos*, Ray, and other botanists ; *Petalodes* of Tournefort. See *Petalodes*.

PERFOLIATUM *Folium* [*per* & *folium*]

S

si basis folii undique cingat transversim caulem, when the *basis* of the leaf entirely surrounds the stem transversely ; differing from *Amplexicaule* in the stem appearing nearer the centre of the leaf, as in the *Bupleurum rotundifolium, Uvularia perfol.*

PERFORATÆ *Cotyledones* [*perforor,* to be pierced through] perforated. A species of the *Monocotyledones,* exemplified in *Gramina.*

Perforatæ, an order of plants in the *Fragmenta methodi naturalis* of Linnæus, containing the following *genera,* viz. *Hypericum, Ascyrum, Cistus, Telephium.*

PERIANTHIUM [ex περι, *circum,* about, & ανθος, *flos,* a flower] the *Calyx* so called when contiguous to the fructification. When it includes the *Stamina* and *Germen,* it is the *Perianthium* of the fructification ; when the *Stamina,* without the *Germen,* are included, the *Perianthium* of the flower ; the *Perianthium* of the fruit, when it contains the *Germen,* and not the *Stamina.*

PERICARPIUM [from περι, *circum,* & καρπος, *semen,* seed] the *Germen* so called in its state of maturity. It is that *Viscus,* as Linnæus terms it, which contains the seed of the plant,
and,

and, in different fubjects, is diftinguifhed by the feveral appellations of *Capfula, Siliqua, Legumen, Conceptaculum, Drupa, Pomum, Bacca, Strobilus, Folliculus.*

PERICHÆTIUM [à περι, & Χαιτη, *juba*] A modification of the *Receptaculum* in the *Mufci* & *Algæ*; it is a cylindrical fheath in the *Polytricum, fquammofum* in the *Hypnum, tubulofum* and ferving for the *Calyx* in the *Jungermannia.*

PERPENDICULARIS *Radix.* According to Linnæus, a fpecies of *Caudex defcendens* which runs ftraight down into the ground.

PERSISTENS *Folium* [*perfifto*, to abide] expreffes the third degree of duration, fee *caducum, deciduum*; remaining till the fruit is ripe, *ad maturitatem fructus perfiftens.*

Perfiftentes Stipulæ, exemplified in the clafs *Diadelphia,* and *Icofandria polygynia.*

Perfiftens Calyx, as in the clafs *Didynamia.*

PERSONATÆ [i. e. *perfonam gerens*; mafked] An order of plants in the *Fragmenta methodi naturalis* of Linnæus, containing thefe *genera,* viz. *Cymbaria, Antirrhinum, Rhinanthus, Pedicularis, Bartfia, Euphrafia, Melampyrum, Obolaria, Orobanche, Jufticia, Verbena, Veronica,* &c. The

Perfonatæ

Perſonatæ are of the third claſs in Tournefort, and are the *Didynamia Angioſpermia* of the ſexual ſyſtem.

Perſonatus. See *Ringens.*

PES, a foot. The eighth degree in the Linnæan ſcale for meaſuring the parts of plants : from the elbow to the *baſis* of the thumb, or twelve Pariſian inches. See *Menſura.*

PETALIFORMIA *Stigmata* [*Petalum*] In ſhape reſembling a *petalum,* as in the *Iris.*

PETALODES *Flos* [*Petalum*] Tournefort. Having a *corolla : Perfeɛtus* of many other botaniſts.

PETALUM [from Πεταω, *pando,* to expand] the corollaceous tegument of the flower. The leaves which cɔnſtitute the *Corolla* are called *Petala,* to diſtinguiſh them from the other leaves of the plant. The different parts of a *Petalum,* or *Petala,* are the *Tubus, Limbus, Unguis,* & *Lamina.*

PETIOLARIS *Cirrus* [*petiolus*] a tendril proceeding from the foot ſtalk of a leaf.

Petiolaris

Petiolaris Pedunculus, a *Pedunculus* inferted in a *Petiolus*, as in the *Hibifcus, Turnera*.

Petiolaris Gemma, regards the origin of the *Gemma* ; formed from a *petiolus :* it is either *oppofitiva*, or *alternativa*.

Petiolares Glandulæ, when the *glandulæ* are on the *petioli*, as in the *Ricinus, Jatropha, Paffiflora, Caffia, Mimofa*, &c.

PETIOLATUM *Folium* [*Petiolus*, a foot-ftalk] a leaf growing on a foot-ftalk, oppofed to *Seffile*.

PETIOLUS [dim. à *pede, quafi pediolus*, a lit · tle foot, vel à *petilus*, flender] the foot-ftalk of a leaf, *folium ; trunci fpecies adneftens folium, nec fruftificationem :* yet there are fome few inftances where the fame foot-ftalk fupports both fruftification and leaves, as in the *Turnera, Hibifcus*. Diftinct from the foot-ftalk of a flower which is called *Pedunculus*. One of the feven *Fulcra* in the Linnæan fyftem.

PILEUS *Fungi* [à Πῖλος, *lana coaɛ́ta*] a hat or bonnet. The orbicular expanfion of a mufhroom, which covers the fruftification.

PILI, hairs ; a fpecies of *pubefcens* which frequently covers the furface of plants, and ferves
them

them as excretory ducts, *ductus excretorius plan-tæ fetaceus.*

PILOSUM *Folium* [Πιλος, *pilus*, a hair] when the furface of the leaf is covered with long ·diftinct hairs, as in the *Cortufa, Juncus pilofus, fylvaticus, campeftris.*

Pilofa Semina, hairy, as in the *Centauriea,* & *Tragopogon.*

PINNATIFIDUM *Folium* [Πιννα, *Pinna*, a wing] applied to a fimple leaf, *tranfverfim divi-fum laciniis horizontalibus oblongis,* tranfverfely divided into long horizontal *laciniæ,* as in the *Caucalis arvenfis,* & *anthrifcus, Chelidonium hybri-dum, Papaver rhoeas, dubium.*

PINNATUM *Folium* [*pinna*, the large fea-thers of a wing] that fpecies of compound leaf in which many *foliola* grow on the fides of one petiolus ; *cum petiolus fimplex lateribus adnectit foliola plura. Folia Pinnata* are either *bijuga, tri-juga, quadrijuga,* &c. ; *impari abrupte, cirrhifa, foliis oppofitis, alternis, interruptis, articule tis, decurfivis.* The *Polemonum cæruleum,* and *Agri-monia eupatoria,* fhew examples of the *folium pinnatum* ; likewife, the *Potentilla fruticofa, ar-gentina, rupeftris.*

PIPERITÆ [*Piper*, pepper] An order of plants in the *Fragmenta methodi naturalis* of Linnæus. The plants of this order are thefe, *A-rum, Dracontium, Calla, Acorus, Saururus, Pothos, Piper, Phytolacca.*

PISTILLUM, a peftil. The little upright column which is generally found in the centre of every flower. According to the Linnæan fyftem, it is the female part of generation, whofe office is to receive and fecrete the *Pollen*, and produce the fruit. It confifts of three parts, viz. *Germen, Stylus*, and *Stigma.*

PIXIDATUM *Folium* [à *Pixis*, a box] a fpecies of the *articulatum*, when one leaf is as it were let into the other, *fi unum folium in foveam alterius quafi immittitur*, Ludw. as in the *Equifetum* & *Ephedra.*

PLACENTA, *Boerhaave.* See *Receptaculum.*

PLACENTATIO [*Placenta*] regards the *Cotyledons* of the feed, *cotyledonum difpofitio fub ipfa feminis germinatione.* Plants, as to placentation, are *acotyledones, monocotyledones, dicotyledones*, or *polycotyledones.*

PLANIPETALUS *Flos* [*planus*, plane, flat,
&

& *petalum*] *Semiflofculofus*, of Tournef. *Lingulatus* of Ponted. *Cichoraccus* of Vaill, See *Ligulatus*.

PLANTÆ, plants ; one of the feven tribes, or families, of the vegetable kingdom, according to Linnæus : it comprehends all vegetables which are not included in the other fix, which fix families are thefe, viz. *Fungi, Algæ, Mufci, Filices, Gramina, Palmæ. Plantæ* are either *herbaceæ, frutices,* or *arbores.*

PLANUM *Folium* [ab ἀπλανης, *planus*] plain, flat, neither, neither convex nor concave, *quod utramque fuperficiem ubique parallelam gerit.*

PLENUS *Flos*, a full· flower, whofe *Corolla* is fo multiplied as to exclude all the *Stamina.* This plenitude is brought about by the *Stamina* running into *Petala*, which happens moft frequently to polypetalous flowers. Thefe *flores pleni,* fays Linnæus, becoming eunuchs, are always barren, and being unnatural productions, conftitute no *genus.* The impletion of *fimple* flowers is by the increafe either of the *Petala* or *Nectarium ;* that of *compound* ones, by the *Radius* or the *difk.*

PLICATUM *Folium* [*plico*, to fold] *quum difcus*

*dlifcus folii verfus marginem ad angulos adfcendit &
defcendit*; in plain Englifh, plaited like a woman's fan, or a candle-fhade, as in the *Eryngium
maritimum*. When ufed as a term of foliation,
it implies the leaf within the *gemma*, being
plaited longitudinally, as in the *Betula*, *Vitis*,
Malva, *Urtica*, *Alchemilla*, *Paffiflora*, *Viburnum*, &c.

Plicatæ Cotyledones, plaited or folded, a fpecies of the *Dicotyledones*, exemplified in the *Goffypium*.

Plicata Corolla, as in the *Convolvulus*.

PLUMATA *feta* [*pluma*, a foft feather]
That fpecies of pubefcence termed *feta*, is faid
to be *plumata*, feathered, when each briftle
has lateral hairs growing like the plume on a
quill.

PLUMOSUS *Pappus* [*pluma*, a fmall foft
feather] compound and woolly, as in the *Crepis*, *Scorzonera*, *Tragopogon*, &c.

PLUMULA, a little feather; the afcending
fcaly part of the *Corculum* of the *Semen*.

POLLEN [from Παλη, fine meal, or flower]
that duft which is contained in the *Anthera*, and
which, according to Linnæus, is difcharged
thence,

thence, and lodges on the surface of the *Stigma*, by whose moisture it is detained; which moisture bursts its atoms and diffolves them, and thence passing through the *Stylus*, impregnates the *Germen* below; which *Germen*, thus impregnated, swells and produces the fruit.

POLLEX, the thumb. The fourth degree in the Linnæan scale for measuring the parts of plants : the length of the first joint of the thumb, or a Parisian inch. See *Menfura*.

POLLICARIS *Caulis* [à *Pollex*, a thumb] See *Pollex*.

POLYADELPHIA [à πολυς, *multus*, many, & Αδιλφος, *frater*, a brother] the eighteenth class in the sexual system of Linnæus; it includes those plants which bear hermaphrodite flowers with three or more sets of united *Stamina*. The orders are three, *viz.* PENTANDRIA, containing but one *genus*, viz. *Theobromo*; ICOSANDRIA, containing but one *genus*, viz. *Citrus*; POLYANDRIA, containing 2 *genera*, viz. *Hypericum, Afcyrum.*

POLYANDRIA [Πολυς, *multus*, many, & Ανηρ, *maritus*, a husband] the thirteenth class in the Linnæan system; it contains those plants

plants which bear hermaphrodite flowers, with many *stamina*, and is to be distinguished from the *Icosandria*, in wanting the classical characteristics of that class. The orders are seven, viz. MONOGYNIA, comprehending 38 *genera*, viz. *Marcgravia, Rheedia, Morisonia, Capparis, Actæa, Sanguinaria, Podophyllum, Chelidonium, Papaver, Argemone, Cambogia, Muntingia, Sarracenia, Nymphæa, Bixa, Sloanea, Mammea, Ochna, Calophyllum, Grias, Tilia, Laetia, Microcos, Elæocarpus, Lecythis, Messua, Vateria, Lagerstroemia, Thea, Caryophyllus, Mentzelia, Plinia, Delima, Cistus, Prockia, Corchorus, Seguiera, Symplocos*; DIGYNIA, containing 3 *genera*, viz. *Pæonia, Curatella, Calligonum*; TRIGYNIA, containing 2 *genera*, viz. *Delphinium, Aconitum*; TETRAGYNIA, containing but one *genus*, viz. *Tetracera*; PENTAGYNIA, containing 3 *genera*, viz. *Nigella, Aquilegia, Reaumuria*; HEXAGYNIA, containing but one *genus*, viz. *Stratioles*; POLYGYNIA, containing 17 *genera*, viz. *Dillenia, Liriodendron, Magnolia, Michelia, Uvaria, Annona, Anemone, Atragene, Clematis, Thalictrum, Adonis, Ranunculus, Trollius, Isopyrum, Helleborus, Caltha, Hydrastis.* Polyandria plerumque venenata est. *Lin.*

POLYANGIÆ [πολυς, many, & Ἀγγ⊙·, *vas*, a vessel, or *loculamentum*] The twentieth class

in

in Boerhaave's syftem ; it contains the *Malva, Nymphæa, Nigella, Ciftus,* &c.

POLYCOTYLEDONES [Πολυς, & *Cotyl.*] having many cotyledons ; a mode of placentation, exemplified in the *Pinus, Cupreffus,* & *Linum.*

POLYGAMIA [πολυς, *multus,* many, & Γα. μος, *nuptiæ,* nuptials] The twenty-third clafs in the Linnæan fyftem, comprehending thofe plants which bear hermaphrodite flowers, together with male or female flowers, or both; *mariti cum uxoribus & innuptis cohabitant in diftin-Etis thalamis.* The orders in this clafs are three, viz. MONOECIA, containing 18 *genera,* viz. *Mufa, Holcus, Cenchrus, Ifchæmum, Ægilops, Andropogon, Apluda, Valantia, Ophioxylon, Celtis, Veratrum, Acer, Begonia, Mimofa, Delechampia, Clufia, Parietaria, Atriplex* ; DIOECIA, containing 8 *genera,* viz. *Panax, Diofpyrus, Nyffa, Flaxinus, Anthrofpermum, Arctopus, Gleditfia, Pifonia* ; TRIOECIA, containing but one *genus,* viz. *Ficus.*

Polygamia, applied to a fingle flower, regards the intercommunication of the *flofculi* which form that flower, as in the firft, fecond, third, and fourth orders of the clafs *Syngenefia.*

fia. See *Æqualis, Superflua, Fruftranea,* &
Neceffaria.

POLYGYNIA [πολυς, *multus,* many, & Γυνη,
mulier, a woman] one of the orders in the 5th,
6th, 12th, & 13th claffes in the Linnæan fyftem : in thefe claffes it diftinguifhes the plants
in whofe fructification there are many *Styli,*
which are confidered, in the fexual fyftem, as
the female organs of generation.

POLYPETALA *Corolla* [πολυς, *multus,* &
πεταλον, *petalum*] confifting of many *petala,* as
in the *Nymphæa.*

POLYPHYLLUM *Involucrum* [πολυς, *multus,*
many, & Φυλλον, *folium,* a leaf] confifting of
many leaves.

POLYSTACHIUS *Culmus* [πολυς, & Σταχυς,
fpica] having many *fpicæ,* as the *Scirpus lacuftris,
Scirpus holofchœnus,* & *Scirpus fetaceus.*

POMACEÆ [*Pomum,* an apple, pear, &*c.*]
An order of plants in the *Fragmenta methodi na-
turalis* of Linnæus, containing thefe *genera,* viz.
*Punica, Pyrus, Cratægus, Mefpilus, Sorbus,
Ribes.*

POMUM, an apple, pear, &c.; according to Linnæus, a close pulpy *Pericarpium*, covered by a continued thin membrane without valves, and containing a *Capsula*.

PORI [πεῖρω, *transadigo*, to pierce through]. Linnæus chuses to clafs thefe *Pori*, pores, obfervable on the furface of fome plants, among the fecretory, rather excretory, *glandulæ*. We have examples of thefe pores in the *Tamarix*, *Silene*, &c.

POSTICUS *Angulus*, [a poft, *ut anticus ab ante*] a pofterior angle, fuch as are formed by the excavation in the *bafis* of a *Folium cordatum*, *lunatum*, or *fagittatum*.

PRÆMORSA *Radix*, [à *præmordeo*, to bite] a root which does not run tapering to its extremity, but appears truncated, or bitten off, as in *Scabiofa*, *Plantago*, & *Valeriana*.

Præmorfum folium, a leaf whofe *apex* is very obtufe, and unequally notched or bitten, *quod obtufiffimum terminatur incifuris inæqualibus*.

PRECIÆ [*precius*, early] an order of plants in the *Fragmenta methodi naturalis* of Linnæus, containing thefe *genera*, viz. *Primula*, *Androface*,

Androſace, Diapenſia, Cortuſa, Dodecatheon, Soldanella, Cyclamen.

PRISMATICUS *Calyx* [*Priſma,* a priſm] equal in diameter from top to bottom, but different from *Cylindraceus,* in its circumference being angular, as in the *Pulmonaria.*

Priſmaticum Pericarpium, cum lineare poly-edrum fit lateribus planis.

PROCUMBENS *Caulis :* lying along the ground, *horizontaliter ſupra terram ;* different in ſignification from *repens,* in not ſhooting out *radiculæ* as it runs along : exemplified in the *Convolvulus ſoldanella.* Synonym. with *pro-ſtratus.*

PROLIFER *Flos* [from *Proles,* offspring]. Flowers are called *proliferous* where one grows out of the other, which ſeldom happens except in *flores pleni.* Prolification is either from the centre or from the ſide : the firſt happens in *ſimple* flowers, when the *Piſtillum* ſhoots in-to another flower raiſed on a ſingle *Pedunculus ;* the ſecond, in aggregate flowers, properly ſo called, when, from one common calyx, many pedunculate flowers are produced. When um-bellate flowers become proliferous, it is by one *Umbellula* growing out of another.

Prolifer

Prolifer Caulis, shooting forth branches only from the centre of the *apex,* as in the *Pinus.*

PROMINULUM *Dissepimentum* [*promineo,* to jet or stand out] prominent at the *apex* of the *Pericarpium* beyond the valves, as in the class *Tetradinamia* of Linnæus.

PRONUM *discum folii* [προνὸς, *antiq.* having the face downwards] the inferior disk or back of the leaf.

PROPAGO, a shoot or layer ; the seed of mosses, first discovered by Linnæus in the year 1750.

PROPRIUM *Involucrum,* when at the *basis* of a *flos umbellatus* properly so called.

PROSTRATUS *Caulis.* See *Procumbens.*

PRUNUS. See *Drupa.*

PSEUDO [ψευδω, *fallo,* to deceive] bastard ; as *Pseudo-cyperus,* bastard cyperus : synon. with *Adulterinus.*

PUBES, down, hair. One of the seven
kinds

kinds of *Fulcra*; it includes *pili, lana, barba, tomentum, ftrigæ, fetæ, hami, glochides, glandulæ, utriculi, vifcofitas, glutinofitas,* In the *Phil. Botanica, ftimuli, aculei, furcæ, fpinæ,* were alfo numbered among the *Pubes*; but Linnæus has fince ranged them under *Arma*.

PUBESCENTIA *eft armatura plantæ quæ ab externis injuriis defenditur.* See *Pubes*.

PULPOSUM *Folium* [*pulpa*, the pulp, or flefhy part of meat] regards the fubftance of leaves, *quod interne pulpa repletum eft :* this is Linnæus's explanation of the term *Carnofum*, but certainly it is more properly applied in this place. See *Carnofum, Compactum*. It is alfo applied to fruits; a common plum is *pulpofum*, an apple *carnofum*.

PULVERATUM *folium* [à *Pulvis*, powder, duft] covered with a kind of meal or duft, as on the inferior difk of the *Frankenia pulverulenta, Bonus henricus,* a fpecies of *Chænopodium*.

PUNCTATUM *Folium* [*punctum*, a point] *quod punctis excavatis adfperfum eft,* befprinkled with hollow dots or points, as in the *Anthemis maritima.*

PUTAMINEA [*Putamen*, a shell] an order of plants in the *Fragmenta methodi naturalis* of Linnæus, containing these *genera*, viz. *Capparis, Breynia, Morifona, Crativa, Marcgravia.*

Q.

QUADRANGULARE *Folium*; a quadrangular leaf; having four prominent angles in the circumscription of its disk. *Triangulare, Quadrangulare*, &c. expres the figure of a leaf considered in one plane.

QUADRIFIDUM *Folium* [*in quatuor partes fissus*] consisting of four divisions, its sinuses linear, and margins straight. See *Fissum.*

QUADRIJUGUM *Folium* [*quatuor, & jugo,* to yoke] a *folium pinnatum* consisting of four pair of *foliola.*

QUADRILOBUM *Folium* [*quatuor,* four, & Λοβος, the tip of the ear] consisting of four lobes. See *Lobatum.*

QUADRIPARTITUM *Folium* [*quatuor*, & *partitus*, divided] consisting of four divisions, *usque ad basin*, down to the base.

QUA-

QUATERNA *Folia* ; by fours : applied to the *folia verticillata*, fignifying the number of leaves of which each *verticillum* confifts.

QUINA *Folia*; by fives : applied to the *folia verticillata*, indicating the number of leaves of which each *verticillum* confifts.

QUINATUM *Folium* [*quinus*, five] expreffive of the number of *foliola* in a *folium digitatum*.

QUINQUANGULARE *Folium* ; having five prominent angles in the circumfcription of its difk. This term, as alfo *Triangulare*, &c. indicate the figure of a leaf confidered in one plane.

QUINQUEJUGUM *Folium* [*quinque*, & *jugo*, to yoke] a *folium pinnatum* of five pair of *foliola*.

QUINQUELOBUM *Folium* [*quinque*, five, & Λοϐος, the tip of the ear] confifting of five lobes. See *Lobatum*.

QUINQUEPARTITUM *Folium* [*quinque*, & *partitus*, divided] confifting of five divifions, *ufque ad bafin*, down to the bafe.

Quinquepartitus Calyx, as in the *Lithofpermum*.

QUINQUIFIDUM *Folium* [*in quinque partes fiffum*] confifting of five divifions, with linear finufes and ftraight margins. See *Fiffum*.

Quinquifidus

Quinquifidus is applied to a monophyllous *Ca-lyx* with five fegments, which is a claffical cha-racteriftic of the *Didynamia*.

Quinquifida Corolla, as in the *Myofotis fcorpi-oides*.

R.

RACEMUS [à *ramus*, vel à *radendo*] a bunch of grapes or other berries ; a fpecies of inflo-refcence refembling a bunch of currant-berries, confifting of a *pedunculus* with fhort lateral branches, as in the *Vitis, Ribes*, &c. A *Race-mus* may be *fimplex, compofitus, unilateralis, peda-tus, conjugatus, erectus, laxus, dependens, nudus*, or *foliatus*.

RACHIS [Ραχις, *dorfum*, the back ; or ra-ther, *fpina dorfi*, the back-bone] A fpecies of *receptaculum*, as in the *Panicum crocus corvi*, & *crocus galli, Senecio vulgaris*.

RADIATUS *Flos* [*Radius*, a ray] A fpecies of compound flower in which the *Corollulæ* of the *Difcus* are tubular, and thofe of the mar-gin either *ligulatæ, tubulofæ*, or *fubnudæ*. Clafs *Syngenefia* of Linnæus.

RADICALIA *Folia* [*Radix*, a root] a term relative merely to what is called the determination of leaves ; leaves proceeding immediately from the root, as in the *Potentilla opaca*.

Radicalis Pedunculus, a foot-ftalk proceeding from the root.

RADICANS *Caulis* [*radicor*, to take root] bending to the earth, and ftriking root, but not creeping along ; in this refpect different from *Sarmentofus*, & *Repens*.

Radicans Folium, in aquatic plants, when the leaves ftrike root.

RADICATUM *Folium* [*Radix*] fhooting out roots from the fubftance of the leaf.

RADICULA [dim. à *radix*] a little root ; the fibrofe part of the *Radix*, terminating the defcending *caudex*, and imbibing nourifhment for the fupport of the plant.

RADIUS, a ray ; the circumference, or margin, which furrounds the *Difcus* in a radiate compound flower.

RADIX, a root ; commonly underftood to be that part of the plant which is underground ; but Linnæus chufes to confider as root all that
lies

lies below the branches. The ftem, he terms the afcending *Caudex*, and that which is commonly called the body of the root, the defcending *Caudex*. The root therefore, according to this writer, confifts of *Caudex* and *Radicula* : it is compofed of *medulla, lignum, liber,* & *cortex.* Vid. *Phil. Botan. p.* 38.

RAMEA *Folia* [*Ramus*, a branch] regards the determination only ; leaves growing on the branches, oppofed to *Caulina Radicalia.*

Rameus Pedunculus, the foot-ftalk of a flower proceeding from a branch.

RAMOSISSIMUS *Caulis* [*Ramus*, a branch] *ramis multis abfque ordine gravidus,* abounding with branches irregularly difpofed.

RAMOSUS *Caulis* [*Ramus*, a branch] having many branches.

Ramofa Radix, having ftrong lateral branches, as in the *Urtica.*

RAMUS [ab ὄξαμνος, a fmall branch] the branch of a tree.

RECEPTACULUM, a receptacle ; the bafis on which the other fix parts of fruçtification are conneçted : its fpecies are *Receptaculum proprium,*

prium, *Receptaculum commune*, *Umbella*, *Cyma*, *Spadix*.

Receptaculum Commune, common receptacle, connecting many *flosculi*, so as that taking any of them away would cause irregularity.

Receptaculum Floris, receptacle of the flower, a *basis* to which are fixed the parts of the flower exclusive of the *germen*.

Receptaculum Fructificationis, receptacle of the fructification, common to the flower and fruit.

Receptaculum Fructus, receptacle of the fruit, a *basis* for the fruit only, remote from that of the flower.

Receptaculum Proprium, proper receptacle ; belonging to one fructification only.

Receptaculum Seminum, receptacle of the seed, is the *basis* on which the seeds are fixed within the *Pericarpium*.

Receptaculum, *Sedes* of Ray, *Placenta* of Boerhaave, *Thalamus* of Vaill.

RECLINATUM *Folium* [*reclino*, to bend] *quod deorsum curvatur*, bending downward, so that the *apex* of the leaf is lower than the base. The same as *Reflexum*. *Arcuatim versus terram*, says Linnæus, applying this term also to the *Caulis*. As a term of foliation, it implies the leaves within the *gemma* being folded back towards the *petiolus*, *versus petiolum deorsum reflexa*,

flexa, as in the *Aconitum*, *Hepatica*, *Adoxa*, *Po-dophylium*, &c.

RECURVATUM *Folium* [*recurvo*, to bend back] bent downward, in a greater degree than *Reclinatum*, but not so much as *Revolutum*.

REDUCTÆ *Cotyledones*, reduced ; a species of the *monocotyledones*, exemplified in the *Cepa* ; and also of the *Dicotyledones*, exemplified in the *Umbellatæ*.

REFLEXUS *Ramus* [*reflecto*, to bend back] bent back again to the trunk ; or bent in two opposite directions. See *Deflexus*, *Retroflexus*.

Reflexum Perianthium, bent back, as in *Ascle-pias Leontodon taraxacum*.

Reflexum folium, as in the *Euphorbia portlan-dica*.

REGULARIS *Corolla* ; regular, equal in the figure, magnitude, and proportion of its parts, as in the *Phillyrea*, *Ligustrum*, *Syringa*, *Jas-minum*.

REMOTUS *Verticillus* [à *removeo*, to remove] when the *Verticilli* are at a considerable distance from each other, opposed to *contiguus* ; ex-emplified in the *Galeopsis ladanum*.

Remota

Remota Folia, oppofed to *approximata.*

Remoti Pedunculi, oppofed to *conferti.*

RENIFORME *Folium* [*Ren,* a kidney] In fhape refembling a kidney, *fubrotundum, bafi excavatum, angulis defitutum,* as in the *Convolvulus foldonella, Campanula rotundifolia, Saxifraga granulata.*

REPANDUM *Folium* [*re,* & *pando,* to bend] *cujus margo angulis, eifque interjectis finubus, circuli fegmento infcriptis terminatur;* properly fpeaking, having a ferpentine margin; without any angles at all.

REPENS *Radix* [à *repo,* to creep] a creeping root; *i. e.* extending horizontally, and fending forth *Radiculæ* from fpace to fpace, as in *Mentha.*

Repens Caulis, running along the ground, and ftriking root at certain diftances, as in the *Hedera,* & *Bignonia.*

REPTANS *Flagellum* [*repto,* to creep] creeping along the ground, as in the *Fragaria vefca.*

RESTANTES *Pedunculi;* remaining after the fructification has fallen off.

RESUPINATIO *Florum* ; when the *labium superius* of the *corolla* faces the ground, and the *inferius* is turned face upward, as in the *Violæ europeæ*, *Ocymum*, *Ajuga orientalis*, and some species of the *Satyrium*.

RESUPINATUM *Folium*, [*resupino*, to turn upwards] turned upside down.

Resupinata Corolla, as in the *Schrophularia*.

RETROFLEXUS *Ramus* [*retro*, backward, & *flexus*, bent] according to Linnæus, the third degree of curvature ; three times bent ; bent in three different directions. See *Deflexus*, *Reflexus*.

RETROFRACTUS *Pedunculus*, [ex *retro*, backwards, & *frangor*, to be broken] bent or bowed backward towards its insertion.

RETUSUM *Folium* [*retundor*, to be blunted] the natural meaning of this word is the same as *obtusum*; but Linnæus chuses to understand them very differently. See *Obtusum*. *Retusum* he explains thus, *quod terminatur sinu obtuso*, terminating in an obtuse *sinus*. Both these words regard the *apex*. The *Folium retusum* is exemplified in the *Frankenia pulverulenta*.

REVOLUTUM *Folium* [*revolvo*, to roll back] rolled back. It is particularly used by Linnæus as a term of foliation (see *Foliatio*), signifying the lateral margins being rolled spirally backward, as in *Rosmarinus*, *Tucrium marum*, &c.

Revoluta Corolla, rolled back, as in the *Asparagus*, *Medeola*.

RHÆADES [*Rhœas*, the red poppy] An order of plants in the *Fragmenta methodi naturalis* of Linnæus, containing these *genera*, viz. *Papaver*, *Argemone*, *Chelidonium*, *Bocconia*, *Sanguinaria*, *Actæa*, *Podophyllum*.

RHOMBEUM *Folium* [*Rhombus*, a geometrical figure of four equal sides, but not right-angled] a diamond-shaped leaf.

RHOMBOIDEUM *Folium* [*Romboides*, a geometrical figure whose sides and angles are unequal] exemplified in the *Chenopodium viride*.

RIGIDUS *Caulis*, *folia* ; stiff, opposed to *laxus.*

RIMOSUS *Caulis*, abounding with clefts and chinks.

RINGENS [from 'Ρίν, *nasus*, a nose] grinning. Applied to the irregular division of the *Limbus* of a monopetalous *Corolla* into two lips: *Labiatus*, & *Personatus* of Tournefort; *Monopetala irregularis* of Rivinus : such, in general, are the plants of the class *Didynamia* of Linnæus.

ROSACEUS *Flos* [*Rosa*, a rose] consists of more or less than four *Petala*, placed in a circle, like those of the rose, as in the *Ranunculus, Quinquefolium, Pæonia* : Tournef. class vi.

ROSTELLUM, a little beak ; the descending plain part of the *Corculum* of the *Semen*.

ROTACEÆ [*Rota*, a wheel] An order of plants in the *Fragmenta methodi naturalis* of Linnæus, in which are these *genera*, viz. *Gentiana, Exacum, Chironia, Swertia, Lysimachia, Anagallis, Trientalis*, &c.

ROTATUS *Limbus Corollæ* [*Rota*, a wheel] expanded horizontally, without a tubular *basis*, as in the *Borago, Lysimachia*.

ROTUNDATUM *Folium* ; rounded ; *quod angulis privatur.*

RUBRA *Lactescentia*, red, as in the *Rumex sanguinea*.

RU-

RUDERATIS, *locis* underflood [*Rudus,* rubbifh] growing among rubbifh and in high-ways. *Ruderata,* fays Linnæus, *juxta domos, habitacula, vias, ac plateas.*

RUGOSUM *Folium* [*ruga,* a wrinkle] wrinkled, *cum venæ foliorum contractiores eva-dunt quam difcus ut interjecta fubftantia adfcen-dat,* when, from the contraction of the veins, the fubftance of the leaf rifes above them, as in the *Salvia, Primula vulgaris,* & *veris.*

S.

SAGITTATUM *Folium* [*Sagitta,* an arrow] a leaf fhaped like the head of an arrow, *tri-angulare, bafi excavatum, angulis pofticis inftru-ctum,* as in the *Convolvulus arvenfis,* & *fepium, Rumex acetofa, Erica vulgaris.*

SARMENTACEÆ [*Sarmentum,* a twig or fpray of a vine] An order of plants in the *Frag-menta methodi naturalis* of Linnæus, amongft which are thefe *genera,* viz. *Ciffus, Aitis, He-dra, Panax, Aralia, Rufcus, Afparagus, Uvu-laria, Convallaria, Gloriofa.*

SARMENTOSUS *Caulis* [à *Sarmentum,* the twig

twig of a vine] *repens*, *subnudus*, creeping, almost naked ; producing only a few leaves in bunches juſt above each knot of *radiculæ*, which ſhoot into the ground at various diſtances. When it puts out roots the whole length of the ſtalks, ſays Miller. Producing runners, ſays Hill, as in the *Aſarabacca*.

SCABER *Caulis*, *Folium* ; ſcabby, rough with tubercles ; oppoſed to *Glaber*.

SCABRIDÆ [à *ſcaber*, rough, rugged] An order of plants in the *Fragmenta methodi naturalis* of Linnæus, conſiſting of the following *genera*, viz. *Ficus*, *Dorſtenia*, *Parietaria*, *Urtica*, *Cannabis*, *Acnida*, *Humulus*, *Morus*.

. SCABRITIES [à *ſcaber*, rough] a ſpecies of *Pubeſcentia*, according to the *Phil. Botan.* compoſed of particles, ſcarce viſible to the naked eye, ſprinkled upon the ſurface of the plant. Guettardus, ſays Linnæus, was among the firſt who, *lynceis oculis*, obſerved this particular. *Scabrities* is either *glanduloſa*, *ſetacea*, or *articulata*.

SCANDENS *Caulis* ; climbing, as in the *Hedera*, *Lonicera*.
Scandens foliolum, as in the *Clematis vitalba*.

SCA-

SCAPUS [à Σκάπω, *innitor*, to lean upon] that species of *Truncus*, or stem, which elevates the fructification and not the leaves, as in *Narcissus*, *Hyacinthus*, &c. A simple stalk rising directly from the root, says Hill.

SCARROSUM *Folium* [*scarreo*, to be rough] Linnæus ranges this word among those applied to the substance of leaves; what it is intended to signify, I confess I am ignorant. Some are of opinion it means, every nerve of the leaf being visible on the surface. In the last edition of the *Systema naturæ* 1759, we find *Scariosa*, which must be a typographical blunder.

SCITAMINA [an à *situs*, fair, beautiful, or a *scitamentum*, meat of a pleasant taste ?] An order of plants, all exotics, in the *Fragmenta methodi naturalis* of Linnæus : it contains the following *genera*, viz. *Musa*, *Thalia*, *Alpinia*, *Costus*, *Canna*, *Maranta*, *Amomum*, *Curcuma*, *Kæmpferia*.

SCORPIOIDES *Flos* [*Scorpio*, a scorpion] resembling the tail of the scorpion, as in the *Scorpiurus*. Tournef.

SCUTELLATI. See *Scutellum*.

SCU-

SCUTELLUM *Lichenibus* [*Scutum*, a target] A fpecies of fructification which is orbicular, concave, and elevated in the margin, as in fome fpecies of the *Lichen*.

SCYPHIFER [à Σκυφος, *fcyphus*, a cup, & *fero*, to bear] cup-bearing, a fubdivifion of the *genus Lichen*, in Linnæus.

SECRETORIA *Scabrities*, [à *fecerno*, to feparate] a fpecies of glandular *Scabrities*, fcarce vifible to the naked eye, on the furface of fome plants, ferving them as organs of fecretion.

SECUNDA *Spica* [*fequendo*] the flowers turned all one way, *ad unum latus verfis*, as in the *Dactylis cynofurcides*.

Secunda Panicula, as in the *Dactylis glomeratus*, & *Feftuca*.

SECURIFORMIS *Pubefcentia* [*fecuris*, an axe or hatchet] a fpecies of *pubes*, on the furface of fome plants, the *fetæ* refembling an axe, as in the *Humulus*, &c.

SEDES ; Ray. See *Receptaculum*.

SEMEN, feed. Linnæus, in his general definition,

finition, calls it the *deciduous* part of the vegetable, and rudiment of a new production, being vivified by the irrigation or fprinkling of the *Pollen*; but, according to the fame writer, the *Semen*, properly fo called, is the rudiment of a new vegetable, moiftened with juice or fap, and involved in a membrane refembling a bladder. Its principal conftituent parts are *Corculum, Cotyledon, Hilum, Arillus, Coronula.*

SEMINALE *Folium* [*Semen*, feed] feminal leaves, are thofe which before were the *cotyledons*, and appear firft.

SEMITERES *Caulis*; half-cylindrical, flat on one fide, and round on the other.

SEMPERVIRENS *Folium* [*femper*, & *virens*, green] ever-green; the longeft degree of duration. See *Caducum, Deciduum, Perfiftens.*

SENA *Folia* [à *fex*] growing in fixes, as in the *Galium fpurium.*

SENTICOSÆ [*Sentis*, a brier, a bramble] an order of plants in the *Fragmenta methodi naturalis* of Linnæus, containing thefe *genera,* viz. *Rofa, Rubus, Fragaria, Potentilla, Tormen-*
tilla,

tilla, Sibbaldia, Drycas, Geum, Comarum, Aphanes, Alchemilla.

SEPIARIÆ [*Sepes,* a hedge] an order of plants in the *Fragmenta methodi naturalis* of Linnæus, containing these *genera,* viz. *Nyctanthes, Jasminum, Ligustrum, Brunsfelsia, Olea, Chionanthus, Fraxinus, Syringa.*

SEPTUM [à *sepio,* to inclose] Ludw. The *Dissepimentum* of Linnæus, which see.

SERICEUM *Folium* [*sericum,* silk] a term applied to those leaves whose surface is remarkably soft, silky, covered with a down of an extreme fine texture.

SERPYLLIFOLIA [*Serpyllum,* Thyme, & *folium*] with leaves resembling those of the *Thymus* of Linnæus, the *Serpyllum* of every other botanist ; the leaves of which are *plani, obtusi, basi ciliati.*

SERRATUM *Folium* [*Serra,* a saw] *quod angulis acutis imbricatis extremitatem respicientibus notatur,* whose margin is notched with imbricated angles, whose shortest side is next the *apex* ; so that *radii,* to bisect each saliant angle, must

be

be drawn from the *bafis* of the leaf, as in the *Vaccinium myrtillus*, *Arbutus unedo*, & *alpina*.

Serratus Calyx; when the *apex*, or upper edge, is regularly cut in fmall *laciniæ*, as in fome fpecies of the *Hypericum*.

Serrata Corolla, as in the *Tilia*, & *Alifma*.

SESSILE *Folium* [à *fedeo*, to fit] growing immediately on the *caulis*, without any *petiolus*, as in the *Tormentilla erecta*, *Teucrium fcordium*, *Mentha fpicata*, *longifolia*.

Seffilis Flos, having no *pedunculus*.

Seffilis Radix, joined to the ftem, as in the *Canna*: a fpecies of tlic *Tuberofa*, according to Dr Hill.

SETÆ [Χαίτη, *juba*, a horfe's mane] briftles: a fpecies of Pubefcence covering the furface of fome plants. *Setæ* are either *fimplices*, *hamofæ*, *ramofæ*, *plumatæ*, or *ftellatæ*: to which, from the *Phil. Botan.* we may add, *cylindricæ*, *conicæ*, *glanduliferæ*, *furcatæ*, as in the *Lavendula*; or *fecuriformes*; as in the *Humulus*, &c.

SETACEUM *Folium*, [à *Seta*, a briftle] covered with a kind of briftly pubefcence, as in the *Afparagus officinalis*.

SEXUS *Plantarum*. Plants are diftinguifhed

ed by the fex of their flowers, which are either *mafculus*, *femineus*, or *hermaphroditus* ; which fee.

SILICULA [dim. à *Siliqua*, a pod] a fpecies of bivalvular *pericarpium*, whofe tranfverfe diameter is equal, or nearly fo, to its longitudinal ; it forms the firft order of the clafs *Tetradynamia*; is diftinguifhed from the *Siliqua*, by the equality of its tranfverfe and longitudinal diameters ; from the *Legumen*, by its feeds being alternately fixed to oppofite futures.

SILICULOSA [à *Silicula*, a little pod] the firft order in the clafs *Tetradynamia* of Linnæus ; containing thofe plants whofe *pericarpium* is a *Silicula*. Thefe are the *Siliculofæ* of Ray, and, together with the *Siliquofa*, the *Cruciformes* of Tournefort.

SILIQUA, a pod, is that kind of *Pericarpium*, which confifts of two *Valvulæ*, and in which the feeds are fixed alternately to each future, *fecundum futuram utramque*. Miller improperly applies this definition to the *Legumen*.

SILIQUOSA [*Siliqua*, a pod] the fecond order in the clafs *Tetradynamia* of Linnæus, containing thofe plants whofe *pericarpium* is a *Siliqua*.

Siliqua. Thefe are the *Siliquofæ* of Ray, and part of the *Cruciformes* of Tournefort.

Siliquofæ, an order of plants in the *Fragmenta methodi naturalis* of Linnæus, of which are thefe genera, viz. *Myagrum, Anaftatica, Subularia, Lepidium,* &c. Siliquofæ aquofæ, acres, incidentes, abftergentes, & diureticæ funt :.exficcatione imminuitur virtus. *Lin.*

SIMPLEX *Caulis* ; a fimple ftem ; *continuata ferie verfus apicem extenditur,* i. e. not dividing, but continuing a fingle ftem up to its *apex,* only fending out fmaller branches.

Simplex Spica, confifting of a fingle *Spica,* oppofed to *compofita fpicillis.*

Simplex Fruêtificatio, oppofed to *compofita ex flofculis.*

Simplex Umbella, having no *Umbellulæ* on the *apices* of its *pedunculi.*

Simplex Radix, not fubdivided.

Simplex Folium, oppofed to *compofitum* ; when there is but a fingle leaf on a *petiolus.*

Simplex Calyx, when confifting of one feries of *Laciniæ,* as in the *Tragopogon.*

SIMPLICISSIMUS *Caulis* ; moft fimple ; having very few branches, and proceeding in a ftraight line up to its *apex,* oppofed to *prolifer,*

fer, dichotomus ; exemplified in the *Lathræa squamaria*. See *Simplex, Integer.*

SINUATUM *Folium* [*Sinus*, a hollow] *quod lateribus finus dilatatos admittit*, whofe lateral finufes are much dilated ; gaping wide. In general, any deficiency or break in the difk of a leaf is termed a *finus*.

SITUS *Foliorum* ; the difpofition of leaves on the ftem, viz. *ftellata, terna, oppofita, alterna, fparfa, conferta*, &c.

SOLIDUS *Caulis*, a folid *Caulis*, or ftem, in oppofition to *inanis*, and *fiftulofus*. *Solida*, when applied to *Radix*, indicates a fpecies of the bulbous root, oppofed to *Tunicata* and *Squammofa*, of a folid fubftance like the turnip.

SOLITARIUS *Pedunculus* [à *folus*, alone] when there is but one proceeding from the fame part.

Solitarius flos, when there is but one flower upon each *pedunculus*, as in the *Euphorbia peplis*; oppofed to *bini, terni*, &c.

Solitariæ ftipulæ, as in the *Melianthus*, growing on the infide, and external in the *Rufcus*.

SO-

SOLUTÆ, *Stipulæ* [*folvor*, to be loofed] loofe, as in moft plants, oppofed to *adnatæ*.

SPADICEUS [from *Spadix*] applied to an aggregate flower, whofe *Receptaculum*, common to many *flofculi*, is within a *Spatha*.

SPADIX, the *Receptaculum* of a palm; a *Pedunculus* which proceeds from a *Spatha*. A *Spadix* may be either branched, as in palms; or *fimplex*, as in *Dracontium*, &c.

SPARSI *Rami, Pedunculi, Folia*; fcattered without order : *ubi plures abfque ordine prognafcuntur*, fays Linnæus. With regard to branches, an accurate obferver will find, that, notwithftanding their irregular appearance, they form a fpiral line round the trunk, regularly completing the circle in a determinate number of fteps. The *Folia fparfa* are exemplified in the *Lilium candid. bulbifer. componium*.

SPATHA, [from Σπατος, *corium*, fkin] the *Calyx* is fo called when it opens longitudinally, refembling a fheath, and envelopes a *Spadix*, which properly means the receptacle of a palm; but this term is generally applied to other plants whofe flower-ftalks proceed from a fheath.

fheath, as in the *Narciffus*, &c. A *Spatha* may be *univalvis, bivalvis, dimidiata*.

SPATHACEÆ [*Spatha*, a fheath, in the language of botany] An order of plants in the *Fragmenta methodi naturalis* of Linnæus, containing thefe *genera*, viz. *Leucoium, Galanthus, Narciffus, Pancratium, Amaryllis, Crinum, Hæmanthus*,

SPATULATUM *Folium* [*Spathula*, an inftrument ufed to fpread falve] *cujus figura jubrotunda, bafi anguftiore lineari elongata*, roundifh, but lengthened by a narrower linear bafe : fhaped fomewhat like a battledoor.

SPECIES *Plantarum*, is the third fubdivifion in the Linnæan fyftem, and comprehends all the different forms of plants which are fuppofed to have been originally created. Thefe plants, fays Linnæus, have, by the eftablifhed laws of nature, continued to produce others like themfelves ; therefore the *Species plantarum* comprehends all the different invariable forms of plants which are found at this day upon the face of the earth.

SPICA [Σταχυς, *Æolice* Σταχυς, an ear of corn] a fpecies of inflorefcence, refembling an ear of corn, as in the *Lavendula Spica*. Linnæus defines it thus, *flores seffiles sparsim alterni in pedunculo communi simplici*, alternate seffile flowers on a simple *pedunculus*.

Spica secunda, when the flowers are all turned one way.

Spica disticha, when the flowers look both ways.

Examples of the *Spica* may be seen in the *Phœnix*, *Arum*, *Piper*, *Pothos*, *Acorus*, &c. A *Spica* may be *simplex, composita spicillis, glomerata, ovata, ventricosa, cylindracea, interrupta*.

SPICATA. See *Spica*.

SPICILLA [dim. à *Spica*] a little *Spica*; the minute spicate flower of which the *Spica composita* is composed.

SPICULA *Graminibus* [dim. à *Spica*] A partial *Spica*, otherwise called *Locusta*.

SPINÆ, thorns, rigid prickles; a species of *Arma*, growing on various parts of certain plants for their defence : *Spinæ ramorum arcent pecora*. On the branches we find examples in the *Pyrus*, *Prunus*, *Citrus*, *Hippophaës*, *Gmelina*, *Rhamnus*, *Lycium*, &c. on the leaves, in the *Aloe*, *Agave*, *Yucca*, *Ilex*, *Hippomane*, *Theophra-*

Z   *sta,*

fta, Carlina, &c. on the *Calyx,* in the *Carduus, Cnicus, Centauria, Moluccella, Galeopfis,* &c. on the fruit, in the *Trapa, Tribulus, Murex, Spinacia, Agrcmonia, Datura,* &c.

SPINESCENS *Petiolus, Stipula* [à *Spina,* a thorn] terminating in a ftrong fharp point.

SPINOSUS *Caulis, Folium* [à *Spina,* a thorn] covered with ftrong woody prickles, whofe roots are not fuperficial, but proceeding from the body of the ftem. When applied to a leaf, *Spinofum Folium,* it indicates the margin running out into rigid points or prickles, *quod margine exit in acumina duriora, rigida, pungentia.*

SPIRALES *Cotyledones* [*fpira,* a circle, the coil of a cable, &c.] twifted fpirally; a fpecies of the *Dicotyledones,* exemplified in the *Salfola, Salicornia, Ceratocarpus, Bafella,* and all the *Holeraceæ.*

SPITHAMA, a fpan. The fixth degree in the Linnæan fcale for meafuring the parts of plants: the diftance between the extremity of the thumb and that of the firft finger when extended; or feven Parifian inches. See *Menfura.*

SPI-

SPITHAMEUS *Caulis* [à *fpithama,* a fpan]
See *Spithama.*

SPLENDENTIA *Folia,* fhining.   Ludw.
See *Nitidum.*

SQUAMA, a fcale ; one of the *fquamæ*
which form an *Amentum.*

SQUAMOSA *Radix* [from *fquama,* a fcale]
fcaly : a fpecies of the *bulbofa,* as for example,
the *Lilium,* which is compofed of fcales lying
over each other.
Squamofus Pedunculus, having a fcaly furface.

SQUARROSUM *Folium,* &c. [ισχαρα, *fcar-
ra,* vel *fquarra,* fcurf] rough, fcaly, or fcur-
fy ; applied, as one would imagine, to the fuperficies of a leaf, and yet Linnæus, in his *Delineatio Plantæ,* ranges it with the terms relative
to the finufes of leaves.
Squarrofum Perianthium, rough, fcaly, as
in the *Onopordum acanthium.*

STAMEN, flax, thread. The *flamina* are
thofe upright filaments which, on opening a
flower, we find within the *Corolla* furrounding
the *Piftillum.*   According to Linnæus, they
are the male organs of generation whofe office
is to prepare the *Pollen.*   Each *Stamen* confifts

of

of two diftinct parts, viz. the *Filamentum,* and the *Anthera.*

STAMINEUS *Flos* [*Stamen*] having no *co-rolla*; Ray. *Apetalus,* of Linnæus: *Incomple-tus, Imperfectus, Capillaceus,* of other botanifts.

STATUMINATÆ [*Statumen,* a prop, a fupport] An order of plants in the *Fragmenta methodi naturalis* of Linnæus, containing the following *genera,* viz. *Ulmus, Celtis, Bofea.*

STELLATA *Folia* [*ftella,* a ftar] leaves furrounding the ftem like the *radii* of a circle. The fame as *Verticillata.*

Stellata Seta. That fpecies of *Pubes* termed *Setæ,* is called *Stellata* when there is a little ftar, compofed of fmaller hairs, affixed to the *apex* of each briftle.

Stellatæ Plantæ, one of Ray's claffes, of which the plants are now ranged among the *Tetran-dria monogynia* of Linnæus.

Stellatæ, an order of plants in the *Fragmenta methodi naturalis* of Linnæus, in which are thefe *genera,* viz. *Anthofpermum, Rubia, Aparine, Ga-lium, Valantia, Spermacoce, Houftonia, Cornus, Phyllis,* &c. Diureticæ funt. *Lin.*

STE-

STERILIS *Flos*, barren, Tournef. *Mafculus* of Linnæus.

STIGMA [from Στίζω, *fignum quod inuritur*, a brand] the *apex* or capital of the *Piflilium*, containing the *Vifcus* which receives the *Pollen*. Linnæus compares this organ to the *Vulva* in the female animals.

STIMULI [*σιγμος*, *Siigmulus*, per fync. *Stimulus*] ftings : a fpecies of *Arma* growing upon fome plants for their defence ; *punctura venenata arcent animalia nuda*, as in the *Urtica*, *Jatropha*, *Acalypha*, *Tragia*. Linnæus divides the *ftimuli* into *pungentes* and *urentes*.

STIPATUS *Pappus* [*Stipes*] elevated on *Stipites*.

STIPES [à *συπος*, a ftump] that fpecies of *Truncus*, which is the *bafis* of a *Frons*, and is peculiar to the *Palmes*, *Filices*, and *Fungi*.
Stipes, the thread which elevates and connects the *pappus* with the feed.

STIPULA [à *ftipa*, tow] ftubble. One of the feven *Fulcra* of plants, according to Linnæus : *fquama quæ bafi petiolorum aut pedunculorum*

rum enascentium utrinque adstat ; the small scale
or leaf which grows on each side of the basis of
a young petiolus or pedunculus, as in papilionaceous flowers, *Tamarindus, Cassia, Rosa, Melianthus, Liriodendrum, Armeniaca, Persica, Padus,* &c. The *Stipulæ* are wanting in the *A-
sperifoliæ, Didynamia, Stellatæ, Siliquosæ, Liliaceæ,
Orchideæ,* and in most of the *Compositæ.* *Stipulæ* are either *geminæ, solitariæ, deciduæ, persisten-
tes, adnatæ, solutæ, intrafoliaceæ,* or *extrafoliaceæ.*

STIPULARIS *Gemma* [*Stipula*] formed from
a *Stipula.*
Stipulares Glandulæ, glands produced from
Stipulæ, as in the *Bauhinia, Armeniaca,* &c.

STIPULATIO [à *Stipula*] *est stipularum situs
& structura ad basin foliorum,* the structure and
situation of the *Stipula.*

STIPULATUS *Caulis* [à *Stipula*] applied to
the superficies of a stem, when the outer coat
is stronger, and more brittle, than a membrane ; rather resembling straw. Opposed to
Membranatus. Or possibly, Linnæus may intend that it should signify, bearing *Stipulæ.*

.STO-

STOLO, a fucker or fhoot, as in the *Viola odorata*, & *Ranunculus repens*.

STOLONIFERUS *Truncus, Caulis* [à *Stolo*, a fhoot, or fcion] having fcions or fuckers.

STRIATUS *Caulis, Culmus, Folia* [*Stria*, a flight groove] fuperficially channelled, or fluted, longitudinally, with parallel lines.

STRICTUS *Caulis, Culmus, Folia* [from *ftringo*, to tie faft] *erectiffimus*, perfectly ftraight, ftiff, as the leaves in the *Campanula pa.ula*; oppofed to *laxus, flaccidus.*

STRIGÆ [à *Strigo*, pro *ftringo*, to grafp, to tighten] ridges, rows, ranks : fuch is the meaning of *Strigæ* in its claffical acceptation. *Strigæ*, fays Linnæus, *arcent fetis rigidis animalcula & linguas*, i. e. by their prickles are a defence againft animals and animalcula. What *fetæ rigidæ* have to do with *Strigæ*, is difficult to conceive, unlefs he chufes to call them *Strigæ*, from their ftanding in rows. Now, in the *Delineatio Plantæ*, he has ranged *Strigæ* with the fofter kinds of *pubes*, fuch as *pili, lana, tome.tum.*

STRIGOSUM *Folium* [à *ftrigando*, ftanding ftill,

ftill, *quippe bos præ macie*] lank, lean, or perhaps, drawn up as if hide-bound : at leaft, I
know of no other meaning to this word that
can poffibly be applied to a leaf, unlefs we derive it from *ftriga*, a ridge, but then it will be
difficult to make *ftrigofum* of it.

STROBILUS, a pine-apple ; a *Pericarpium*
formed from an *Amentum*. Linnæus's term for
the *Conus* of other botanifts.

STYLUS [from ςυλος, *columna*, a pillar]
that.part of the *Piftillum* which elevates the *Stigma* from the *Germen*. Linnæus, in his fyftem
of the generation of plants, affimulates this organ to the *Vagina* or *Tubæ Fallopianæ* in the
females of the animal creation.

SUB, in compofition, for *fere*, almoft; *e. g.*
fubcordatum, fubovatum, nearly *cordatum*, nearly *o-
vatum*, &c. ; but when compounded with a term
of number, it fignifies *moft commonly*, as applied to the *Narciffus poeticus, Spatha fubuniflora.*

SUBDIVISUS *Caulis*, a fpecies of the *Caulis
compofitus* in which the branches are irregularly
fubdivided, *fubdivifus in ramos abfque ordine.*

SUBEROSUS *Caulis, folia* [*fub,* & *erodor,*
to

to be eaten into] as if a little eaten, or gnaw-ed. When applied to a leaf, it refpects the margin only, *margine fuberofa*. If I had not found this term applied to leaves, I fhould have been apt to derive it from *Suber*, a cork, and explained it accordingly.

SUBEXCEDENS *Calyx* [*fub*, & *excedo*, to furpafs] exceeding a little the *Corolla* in length, as in the *Milium*.

SUBMERSUM *Folium* [*fubmergo*, to fink under water] applied to aquatic plants : funk below the furface of the water, as in the *Ranunculus aquatilis :* fynon. with *Demerfum*.

SUBRAMOSUS *Caulis* [*fub*, & *Ramus*, a branch] having few branches,

SUBROTUNDUM *Folium* [*fub*, near to, & *rotundum*, round] nearly circular, in circum-fcription.

SUBULATUM *Folium* [*Subula*, an awl] awl-fhaped, *eft inferius lineare, at verfus apicem fenfim adtenuatur* ; linear below, but gradually tapering towards the *apex*, and ending in a point. This term is alfo frequently applied to the *Stamina* of flowers, and is one of the claf-

A a   fical

fical characteriftics of the *Didynamia, filamenta subulata.* The *Arenaria faxatilis,* and *Sedum rupeftre,* fhew examples of the *Folium fubulatum.*

SUCCULENTÆ [*Succus,* juice] An order of plants in the *Fragmenta methodi naturalis* of Linnæus, containing thefe *genera,* viz. *Caĉtus, Mefembryanthemum, Tetragonia, Aizoon, Sempervivum, Sedum, Geranium, Linum, Oxalis, Saxifraga, Fagonia,* &c.

Succulentum Folium, fucculent, regards the fubftance, oppofed to *exfuccum.*

SUFFRUTEX [*fub,* & *Frutex,* a fhrub] An under-fhrub : according to Tournefort, a plant which is perennial, ligneous, not gemmiparous, and in ftature lefs than a *Frutex,* exemplified in the *Lavendula, Thymus,* & *Salvia.*

SUFFRUTICOSUS *Caulis* [*fub,* & *Frutex,* a fhrub]. See *Suffrutex.*

SULCATUS *Caulis, Culmus, Folia* [*Sulcus,* a furrow] deeply channelled or furrowed longitudinally.

SUPERFICIES *Folii* [ex *fuper,* i. e. *fupra,* & *facies,* a face] furface ; *difcum folii fupinum vel pronum tegit,* covers both the fuperior and
inferior

inferior difk. A leaf, with regard to its fuperficies, is *vifcidum, tomentofum, fcabrum, glabrum,
ftriatum,* &c.

SUPERFLUA *Polygamia,* fuprrfluous. The
fecond order in the clafs *Syngenefia* of Linnæus,
comprehending thofe plants in the compofition
of whofe flowers fome of the *flofculi* are hermaphrodite, and others female ; in which cafe, the
fructification being perfect in the hermaphrodites, the females are fuperfluous.

SUPERUS *Flos* [*fuper,* above] when the
receptaculum of the flower ftands above the *Germen* or fruit ; oppofed to *Inferus.*

SUPINUM *difcum folii,* lying with the face
upwards ; the fuperior difk or belly of a leaf.

SUPRA-AXILLARIS *Pedunculus* [*fupra,* a-
bove, & *axilla,* the arm-pit] whofe infertion
is immediately above the *axilla,* formed by a
branch, or leaf, as in the *Afperifoliæ, Potentilla
monfpelienfis.*

SUPRADECOMPOSITA *Folia,* are thofe
Folia Compofita which confift of *foliola* growing
on a fubdivided *petiolus; cum petiolus aliquoties divifus adnectit plurima foliola :* they are of three
A a 2   kinds,

kinds, viz. *tergemina, triternata, tripinnata*. The *Pimpinella glauca* affords an example of the fupra-decompofite leaf ; alfo the *Ranunculus rutæfol.*

SUPRAFOLIACEUS *Pedunculus* [*fupra,* above, & *folium,* a leaf] inferted into the *caulis* above the *petiolus,* juft over the leaf.

SURCULUS, a fcion or twig, exemplified in the *Jungermannia complanata, dilatata,* &c. It feems to be the fame with *Stolo,* which fee.

SYNGENESIA [Συν, *cum,* or *fimul,* together, & Γένεσις, *generatio*] the nineteenth clafs in the fexual fyftem of Linnæus ; it comprehends the flofculofe, or compofite flowers, which are the *Compofiti* of Tournefort, Rivinus, and Ray. It is a natural clafs, if we except the laft order. The claffical characteriftics of the flofculofe flower are as follows : CALYX, a common *Perianthium,* containing the *Receptaculum* and the *Flofculi* ; it contracts when the florefcence is paft, but expands and turns back when the feeds are mature : it is either *fimplex, imbricatus,* or *auctus.* RECEPTACULUM, common to the fructification, receives many feffile *flofculi* on its *difcus,* which is either concave, plane, convex, pyramidal, or globofe, and its furface either

ther

ther naked, villofe, or paleaceous.—The claffical character of the *Flofculi* is, CALYX, a fmall
Perianthium, frequently quinquedentate, perfifting, fixed on the *apex* of the *Germen*, and
becoming the *Corona* of the feed. COROLLA,
monopetalous, with a very narrow, long tube,
fixed on the *Germen*; it is either tubulate, with
the limb campanulate and quinquefid, and the
Laciniæ fpreading and reflexed ; or ligulate,
with the limb linear, plane, turned outwards,
the *apex* entire, tridentate, or quinquedentate,
truncated ; or wanting, having no limb, and
frequently no tube. STAMINA, *Filamenta* five,
capillary, very fhort, inferted in the neck of
the *corollulæ*; *Antheræ* five, linear, erect, forming by their union a tubulated cylinder, quinquedentate, of the fame length of the limb,
PISTILLUM, *Germen* oblong, under the *Receptaculum* of the flower; *Stylus* filiform, erect,
of the length of the *Stamina* perforating the
Cylinder of the *Antheræ*; *Stigma* bipartite, the
laciniæ revolute, fpreading. PERICARPIUM,
no real one, yet in fome inftances a coriaceous
cruft. SEED, one, oblong, often tetragonous,
frequently narrower at the bafe ; if, inftead of
a *Perianthium*, it be crowned with a *Pappus*, it
will be found to confift of many *radii* in a circle, which are either fimple, radiate, or ra-
mofe : this *Pappus* is either feffile or fixed upon

a

a *stipes*. The essential characteristic of a floscu-
lose flower, is the *Antheræ* being united in a cylinder, and a single seed below the *Receptaculum*
of the *flosculi*. The orders are five, *viz*. Po-
LYGAMIA ÆQUALIS, including those plants
which have compound flowers, the *flosculi* being
all hermaphrodite. It contains 37 *genera*, viz.
*Tragopogon, Scorzonera, Picris, Sonchus, Lactuca,
Chondrilla, Prenanthes, Leontodon, Hieracium,
Crepis, Andryala, Hyoseris, Hypochæris, Lp-
sana, Catananche, Cichorium, Scolymus, Elephanto-
pus, Echinops, Arctium, Serratula, Carduus, Cnicus, Onopordum, Cynara, Carlina, Atractylis, Car-
thamus, Stœbe, Bidens, Cacalia, Eupatorium, Ageratum, Stæhilina, Chrysocoma, Tarchonanthus,
Santolina*. POLYGAMIA SUPERFLUA, including those plants which have the *Flosculi* of their
disk hermaphrodite, and those of the *Radius*
female : it contains 31 *genera*, viz. *Tanacetum,
Artemisia, Gnaphalium, Xeranthemum, Carpesium,
Baccharis, Conyza, Erigeron, Tussilago, Senecio,
Aster, Solidago, Inula, Arnica, Doronicum, Hele-
nium, Bellis, Tagetes, Zinnia, Pectis, Chrysanthe-
mum, Matricaria, Cotula, Anacyclus, Anthemis, A-
chillea, Tridax, Amellus, Sigebeckia, Verbesina,
Tetragonotheca, Buphthalmum*. POLYGAMIA
FRUSTRANEA, comprehending those plants
which have the *Flosculi* of their disk hermapro-
dite, and those of the *radius* neuter : this order
contains

contains 6 *genera*, which are all radiate, viz. *Helianthus, Rudbeckia, Coreopfis, Gorteria, Centaurea, Gundelia.* POLYGAMIA NECESSARIA, including thofe plants which have the *Flofculi* of their difk male, and thofe of the *radius* female : this order contains 13 *genera*, viz. *Silphium, Chryfogonum, Melampodium, Calendula, Arctotis, Ofteofpermum, Othonna, Polymnia, Eriocephalus, Filago, Micropus, Sphæranthus, Milleria.* MONOGAMIA, including thofe plants which have fimple flowers : it contains 6 *genera*, viz. *Seriphium, Corymbium, Jafione, Lobelia, Viola, Impatiens.* Syngenefia compofitorum in medicina receptiffima, communiter amara eft. *Lin.*

T.

TEGMENTUM, a cover. The teguments of a flower are the *Perianthium* & *Corolla.*

TERES *Caulis, Folium,* cylindrical : when applied to leaves, it muft be underftood partially of fuch as by their internal pulp are rendered in part cylindrical, *quod maxima ex parte cylindricum eft,* as in the *Allium vineale,* & *oleraceum.*

TER-

TERGEMINUM *Folium compositum* [*ter,* thrice, & *geminus,* double] a species of what Linnæus calls *supra-decomposita;* three times double, *i. e.* when a dichotomous *petiolus* is subdivided, having two *foliola* on the extremity of each subdivision.

TERMINALIS *Flos;* terminating a branch or stem, as in the *Mentha piperita.*

TERNA *Folia;* three and three: applied to the *folia verticillata,* expressing the number of leaves of which each *verticillum* consists.
Terni pedunculi, three proceeding from the same *axilla,* as in the *Impatiens zeyl.*

TERNATUM *Folium* [*ternus,* three] applied to a *folium digitatum* or *pinnatum,* indicating its number of *foliola,* as in the *Rubus cæsius, fruticosus, saxatilis.*

TESTICULATA *Radix.* See *Duplicata.*

TESSELLATUM *Folium* [*Tessella,* the square pieces of wood or stone used in making checkered work] regards only *folia colorata;* checkered.

TETRADYNAMIA [Τισσαρες, *quatuor,* & δύναμις,

Δύναμις, *potentia,* power] the fifteenth clafs in the Linnæan fyftem; it comprehends all the plants which bear hermaphrodite flowers with four long and two fhort *flamina;* befides which it has the following claflical characteriftcs, *viz.* CALYX, *Perianthium* tetraphyllous, oblong, the *foliola* of which are ovato-oblong, concave, obtufe, connivent, gibbous at the *bafis,* the oppo- fite ones equal and deciduous; COROLLA, cru- ciform, four equal *Petala*; *Ungues plano fubula- ti,* erect, *Limbi* flat, and enlarging outwards, obtufe, and hardly touching each other; the *Petala* inferted in the fame circle with the *Sta- mina.* STAMINA fix, erect; two oppofite to each other, about the height of the *Calyx*; four fomewhat longer, but not fo long at the *Corol- la*; *Antheræ* fomewhat oblong, acuminate, fwel- ling at their *bafis,* erect, with their *apices* incli- ning outwards. Clofe to the *bafis* of the two fhorter *Stamina* there is a nectariferous gland, which differs in its appearance in the different *genera*; to avoid compreffing this gland, thefe two filaments make a curve, which renders them fhorter than the other four. PISTILLUM, the *Germen* above the *Receptaculum* growing dai- ly longer; *Stylus,* the length of the longer *fta- mina,* or entirely wanting; *Stigma* obtufe. PE- RICARPIUM, *Siliqua,* bivalvate; frequently bi- locular, opening from the *bafis* to the *apex*;

B b   *Diffepimentum*

Diſſepimentum prominent at the *apex* beyond the *Valvæ*, the prominent part having before ſerved as a *Stylus*. SEMINA, roundiſh, nodding, alternately and longitudinally ſunk in the *Diſſepimentum*; *Receptaculum* linear, ſurrounding the *Diſſepimentum*, and lodged in the ſutures of the *Pericarpium*. This is really a natural claſs, and has been univerſally conſidered as ſuch by ſyſtematic botaniſts. It is the *Cruciformes* of Tournefort, and the *Siliculoſæ* & *Siliquoſæ* of Ray. The plants of this claſs are univerſally eſteemed antiſcorbutic. The eſſential generical charaéteriſtic is commonly to be found in the ſituation of the neétariferous gland. The orders are two, *viz.* SILICULOSÆ, containing thirteen *genera*, viz. *Myagrum, Vella, Anaſtatica, Subularia, Draba, Lepidium, Thlaſpi, Cochlearia, Iberis, Alyſſum, Clypeola, Biſcutella, Lunaria*; SILIQUOSÆ, containing 15 *genera*, viz. *Dentaria, Cardamine, Siſymbrium, Eryſimum, Cheiranthus, Heſperis, Arabis, Turritis, Braſſica, Sinapis, Raphanus, Bunias, Iſatis, Crambe, Cleome*.

TETRAGONUS *Caulis*; four-cornered. See *Trigonus*.

Tetragona Siliqua, as in the *Sinapis nigra*.

TETRAGYNIA [τισσαϱις, quatuor, & Γυνη, mulier, a woman] one of the orders in the 4th,

5th,

5th, 6th, 8th, and 13th claſſes in the Linnæan ſyſtem ; it diſtinguiſhes the plants, in thoſe claſſes, which in their fructification diſcover four *Piſtilla*, theſe being conſidered as the fe-male organs of generation.

TETRANDRIA [from Τεσσαρες, *quatuor*, four, & Ανηρ, *maritus*, a huſband] Linnæus's fourth claſs, comprehending hermaphrodite flowers, with four *ſtamina* of equal lengths. The orders of this claſs are three, *viz.* Mo-nogynia, containing 56 *genera*, viz. *Leucoden-dron, Protea, Cephalanthus, Globularia, Dipſacus, Scabioſa, Knautia, Ailionia, Hedyotis, Spermacoce, Sherardea, Aſperula, Diodia, Knoxia, Houſtonia, Galium, Crucianella, Rubia, Fuchſia, Siphonanthus, Cateſbœa, Ixora, Pavetta, Peteſia, Mitchella, Cal-licarpa, Aquartia, Polypremum, Penœa, Blœria, Buddleja, Exacum, Plantago, Scoparia, Rhacoma, Centunculus, Sanguiſorba, Ciſſus, Epimedium, Cor-nus, Fagara, Tomex, Ptelea, Ludvigia, Oldenlandia, Ammannia, Iſnardia, Trapa, Dorſtenia, Elæagnus, Brabejum, Krameria, Rivina, Salvadora, Cam-phoroſma, Alchemilla* ; Digynia, containing 6 *genera*, viz. *Aphanes, Crucita, Bufonia, Hamame-lis, Cuſcuta, Hypecoum* ; Tetragynia, contain-ing 6 *genera*, viz. *Ilex, Coldenia, Potamogeton, Ruppia, Sagina, Tillæa.*

TE-

TETRANGIÆ [Τεσσαρες, four, & Αγγ☉, *vas*, a veffel, or *loculamentum*] The eighteenth clafs in Boerhaave's fyftem, containing only *Ruta, Peganum*, & *Stramonicum*.

TETRAPETALA *Corolla* [Τεσσαρες, quatuor, & Πεταλον, *petalum*] confifting of four *petala*, as in the clafs *Tetradynamia*.

TETRAPHYLLUS *Calyx* [Τεσσαρες, & Φυλλον, *folium*, a leaf] confifting of four leaves, as in *Sagina, Epimedium*, and in the clafs *Tetradynamia*.

TETRASPERMA *Planta* [Τεσσαρες, & *fperma*, feed] producing four feeds, as the *Afperifoliæ*, & *Verticillatæ*.

THALAMUS [Θαλαμ☉, a bed, a chamber] the *Receptaculum*, fo called by *Vaillant, Ludwig*, &c.

THECA, a bag or fheath, the fame as *Veficula*, in Linnæus ; exemplified in the *Jungermannia furcata*. Dellen.

THYRSATIM. Towered, in a fort of fpike like a pine cone. *Hill.* See *Thyrfus*.

TO-

TOMENTOSUS *Caulis, Folia* [*Tomentum,*
fhort wool, fhorn off] *villis intertextis vix con-*
fpicuis tegitur, ergo fæpius albidum, uti plantæ ma-
rinæ & campeftres ventis expofitæ, covered with a
whitifh down, whofe hairs are interwoven,
and hardly diftinguifhable; exemplified in the
Ceraftium tomentofum, and in the leaves of the
Origanum onites, & *Althæa officinalis.*

TOMENTUM [Τομιον, *fruftum,* a frag-
ment, *feu quod fectione eft ablatum*] fhort wool,
flocks: a fpecies of pubefcence, which covers
the furface of many plants, defending them, in
fome degree, from the violence of the wind;
gaudet fæpius colore incano, as in the *Tomax, Me-*
dicago, Halimus.

TOROSUM *Pericarpium* [à *Torus,* a protu-
berance or fwelling, as of the veins] bunched
out in knobs by the inclofed feed.

TORTA *Corolla* [*torqueo,* to writhe, torture]
twifted, as in the *Nerium, Afclepias, Vinca.* See
Contorti.

TORTILIS *Arifta,* [à *torqueo,* to writhe,
or twift] twifted.

TORULOSA *Siliqua* [à *torulus*, dim. à *torus*] See *Torofum*.

TRANSVERSUM *Diffepimentum*, at right angles with the fides of the *Pericarpium*, oppo- fed to *parallelum*.

TRAPEZIFORMIUM *Folium* [*Trapezium*, a geometrical quadrangle, whofe fides are nei- ther equal nor oppofite] a leaf in the form of a trapezium.

TRIANDRIA [Τρεῖς, *tres*, three, & Ἀνὴρ, *ma- ritus*, a hufband] Linnæus's third clafs, confift- ing of thofe plants which produce hermaphro- dite flowers, with three *Stamina*. The orders are three, *viz*. MONOGYNIA, of which there are 27 *genera*, viz. *Valeriana, Olax, Tamarindus, Rumphia, Cneorum, Camocladia, Melothria, Orte- gia, Loefiingia, Polycnemum, Hippocratea, Cro- cus, Ixia, Gladiolus, Antholyza, Iris, Moræa, Wa- chendorfia, Commelina, Califia, Xyris, Schoenus, Cy- perus, Scirpus, Eriophorum, Lygeum, Nardus*; DI- GYNIA, containing 29 *genera*, viz. *Bobartia, Cornucopiæ, Saccharum, Phalaris, Pafpalum, Pani- cum, Phleum, Alopecurus, Milium, Agroftis, Aira, Melica, Poa, Briza, Uniola, Daßylis, Cynofurus, Feftuca, Bromus, Stipa, Avena, Lagurus, Arundo, Ariftida, Lolium, Elymus, Secale, Hordeum, Tri- ticum;*

ticum; TRIGYNIA, comprehending 10 *genera*, viz. *Eriocaulon, Montia, Proferpinaca, Triplaris, Holofteum, Polycarpon, Mollugo, Minuartia, Queria, Lechea.*

TRIANGIÆ [Τρεις, three, & Αγ[Θ, *vas*, a veffel] the feventeenth clafs in Boerhaave's fyftem, founded on the number of *loculamenta* in the *pericarpium.*

TRIANGULARE *Folium*; a triangular leaf, *cum tres anguli prominentes ambiunt difcum. Triangulare* is ufed to exprefs the figure of a leaf confidered in one plane, and is therefore different in its fignification from *Triqueter*, or *Trigonus*, which fee.

TRICOCCA *Capfula* [Κόκκ Θ·, *granum*, a grain] *trilocularis* with a fingle feed in each *loculamentum.*

Tricocca, an order of plants in the *Fragmenta methodi naturalis* of Linnæus, in which are thefe genera, viz. *Kambogia, Euphorbia, Dalechampia, Clutia, Andrachne, Phyllantus, Ofyris, Croton, Trajia,* &c.

TRICUSPIDATA *Stamina*, three-pointed, as in the *Allium ampeloprafum, arenarium, fphærocephalon.* See *Cufpidatum.*

TRI-

TRIFIDUM *Folium* [*in tres partes fiffus*] thrice divided by linear finufes, having ftraight margins, exemplified in the *Refida lutea*. See *Fiffus*.

TRIFLORUS *Pedunculus* [*tres*, & *flos*, a flower] bearing three flowers : producing three fructifications on each *Pedunculus*.

TRIGONUS *Caulis* [τρεις, *ter*, & γωνια, an-*gulus*, an angle] Linnæus, in the *Phil. Botan.* explains this term, according to its derivation, as fignifying the number of angles ; but in his *Delineatio Plantæ*, a later work, in a fubdivifion, under *figura*, he ufes *angulatus* to exprefs the angles, and in the next fubdivifion he has 3-*queter*, & 3-*gonus* : now 3-*queter* he explains to mean, three plain fides ; therefore *trigonus* muft fignify a triangular figure whofe fides are either concave or convex.

TRIGYNIA [Τρεῖς, *tres*, three, & Γυνη, *mu-lier*, a woman] the third order in the firft thir-teen claffes, except the firft, the fourth, and the feventh, in the Linnæan fyftem : it includes thofe plants which in their fructification dif-cover thre e*Styli*, which are confidered, in the fexual fyftem, as the female organs of gene-ration.

TRIHILATÆ [*Hilum*, which fee] an order of plants in the *Fragmenta methodi naturalis* of Linnæus, in which are thefe *genera*, viz. *Cardiofpermum, Paullinia, Sapindus, Staphylæa, Banifteria, Berberis*, &c.

TRIJUGUM *Folium* [*tres*, & *jugo*, to yoke] a pinnate leaf, of three pair of *foliola*.

TRILOBUM *Folium* [*tres*, three, & Λοβος, the tip of the ear] confifting of three lobes, as in the *Leonurus cardiaca*. See *Lobatum*.

TRINERVIUM *Folium* [*tres*, three, & *nervus*, a nerve or ftring] having three apparent veffels running from the *bafis* to the *apex* without branching or anaftomofing. This term muft not be confounded with *Triplinervium*, which fee.

TRIOECIA [τρεῖς, *tres*, three, & Οικος, *domus*, a houfe] the third order in the clafs *Polygamia* of Linnæus. There is but one genus of the order, *viz.* the *Ficus*, in which there are male, female, and hermaphrodite flowers produced feparately on different plants.

TRIPARTITUM *Folium* [*tres*, & *partitus*, divided] confifting of three divifions, *ufque ad bafin*, down to the bafe, as in the *Eryngium campeftre*.

TRI-

TRIPETALA *Corolla* [τρεις, *tres*, & Πιταλον, *petalum*] confisting of three *petala*, as in the *Alisma*, & *Sagittaria*.

TRIPETALOIDEÆ [*tres*, & *Petalum*] An order of plants in the *Fragmenta methodi naturalis* of Linnæus, containing the following *genera*, viz. *Butomus*, *Alisma*, *Sagittaria*.

TRIPHYLLUS *Calyx* [τρεις, & Φυλλον, *folium*, a leaf] confisting of three leaves, as in *Tradescantia*.

TRIPINNATUM *Folium compositum* [*tres*, three, & *pinnatus*, winged] *triplicato-pinnatum* ; when a *petiolus* is first pinnated by smaller *petioli*, and those by still smaller, which are themselves pinnated with *foliola* ; *cum petiolus adfigit plura foliola bipinnata :* a species of the *supra-decomposita*, according to Linnæus.

TRIPLINERVIUM *Folium* [*triplex*, triple, & *nervus*, a nerve or string] whose apparent vessels are threefold, that is, running by three's from the *basis* to the *apex*, without branching or anastomosing : different in its fignification from *Trinervium*, which see.

TRIQUETRUM *Folium*, *Caulis* [*qu. triquadrus*, i. e. *quadratus in tres angulos*] having three plain sides, *cujus tria latera longitudinalia plana sunt.*

funt. This term, when applied to leaves, regards their substance, as *Triangulare*, their figure considered in one plane. Examples of the *Caulis triqueter*, may be seen in the *Viola tricolor*.

Triqueter culmus, exemplified in the *Carex cespitosa*.

TRISPERMA [*tres*, & *sperma*, seed] producing three seeds, as *Euphorbia*.

TRITERNATUM *Folium compositum* [*tres*, three, & *ternus*, threefold] *triplicato-ternatum*; when the divisions of a triple *petiolus* are subdivided into three's, each subdivision having three *foliola* on its extremity, *cum petiolus adsigit tria foliola biternata*. This Linnæus calls a species of the *supra-decomposita*.

TRIVALVE *Pericarpium* [*tres*, & *valvæ*, doors or valves] consisting of three valves, as in the *Viola, Polemonium, Helianthemum*.

TRUNCATUM *Folium* [*Truncus*, a stump] having its *apex* truncated, or cut off, *quod linea transversali definit*.

TRUNCUS, in general, the body, stem, or stock of a tree or plant; defined by Linnæus

to

to be that which produces the leaves and fructification ; its species, according to his *Phil. Botan.* are seven, viz. *Caulis, Culmus, Scapus, Pedunculus, Petiolus, Frons, Stipes* ; but, in his *Delineatio plantæ,* the three first and the last only
are retained, the 4th, 5th, and 6th being re-
jected. Former botanists applied the word
Truncus, to trees only.

TUBERCULATUS [à *Tuberculum,* a little
pimple or tubercle] having tubercles, as in the
Lichen scriptus, &c.

TUBERCULUM, a little pimple, exemplified in the *Lichen calcareus.*

TUBEROSA *Radix* [à *Tuber,* a knob] a
knobbed root ; *i. e.* consisting of subrotund
bodies collected into a bundle, as in *Pæonia,
Hemerocallis, Solanum,* &c. The *Tuberosa* are
of three kinds, viz. *palmata, fasciculata, & pendula.*

TUBULATUM *Perianthium* ; tubular, as
in the class *Didynamia* of Linnæus ; opposed to
patens, reflexum, inflatum.
    Tubulata Corolla, as in the class *Didynamia* of
Linnæus.

TUBULOSI *Flosculi* [à *Tuba*, a tube] when they are tubular and nearly equal : *Flosculi* of this structure form one of the three divisions of the compound flowers ; they constitute the *flosculosi* of Tournefort.

Tubulosum Perianthium, tubular, opposed to *patens*.

Tubulosum Folium, when, being cut transversely, it appears hollow within.

TUBUS, a tube ; the inferior narrow part of a monopetalous *Corolla*.

THYRSUS, a spear wrapt about with ivy which the ancients carried in their hands at the feasts of Bacchus. Linnæus makes it a species of inflorescence, which may be either *nudus* or *foliatus*. In the *Phil. Botan.* he defines it to be a *panicula coarctata in formam ovatam*, instanced in the *Syringa & Petasites*.

TUNICATUS *Caulis*, *Radix* [*Tunica*, a coat] wrapt in, or consisting of many coats. When applied to a root, it indicates a species of the *Bulbosa*, consisting of concentric layers as in the *Cepa*, for instance, opposed to *Squamosa*, *Solida*.

TURBINATUM *Pericarpium* [*turbo*, a top] broad

broad at the *apex*, and narrow at the *bafis*, as the *Pyrus.*

Turbinatum Perianthium, as in the *Griflea, Memecylon.*

TURGIDUM *Legumen*, fwollen, as in the *Ononis.*

TURIO [à *Tyro*, a novice] the *Gemma* fo called, by Ludwig, when proceeding from the root.

V.

VAGÆ [*vagor*, to wander] the laft order of plants in the *Fragmenta methodi naturalis* of Linnæus, confifting of thofe *genera*, which he could not with propriety range in any of the other natural orders, viz. *Pinguicola, Collinfonia, Buffonia, Hirtella, Montia,* &c.

VAGINALES [*Vagina*, a fheath] an order of plants in the *Fragmenta methodi naturalis* of Linnæus, containing the following *genera*, viz. *Laurus, Helxine, Polygonum, Biftorta, Perficaria, Atraphaxis, Rheum, Rumex.*

VAGINANS *Folium* [*Vagina*, a fheath] the
bafis

bafis of the leaf infolding the ftem, as in a fheath.

VAGINATUS *Caulis, Culmus [Vagina, a* fheath] when they are fheathed by the *bafis* of their leaves, as in the *Polygonum amphibium*, and all the *Gramina*.

VALVULA, a valve ; the pieces of the external fubftance which, in that fpecies of *pericarpium* termed *Capfula*, inclofes the feed or fruit ; *paries quo fructus tegitur externe.*

VARIETAS, variety ; the fourth fubdivifion in the Linnæan fyftem ; it comprehends the various appearances obfervable in plants produced from the fame kind of feed. The caufes of this variety are the differences of climate, fituation, or foil ; and the mode of their appearance is either in magnitude, plenitude, fhape, colour, tafte, or fmell.

VASA [à *vefcendo*, to be eaten, *quod in ea vefcæ ponantur*] veffels. Vegetables are compofed of at leaft three fpecies of veffels, *viz. Vafa fuccofa*, which convey their juices ; *Utriculi*, which preferve them ; and *Trachea*, which attract the air, like the lungs of animals.

VE-

VEGETABILIA [à *vegeto*, to quicken] one of the three kingdoms of nature according to Linnæus, comprehending seven diſtinct families, viz. *Plantæ, Palmæ, Gramina, Filices, Muſci, Algæ, Fungi.*

VENOSUM *Folium* [*Vena*, a vein] *cum vaſa diſcurrentia evadunt ramoſiſſima, & anaſtomoſes nudo oculo exhibent*, whoſe veins branch and anaſtomoſe over the whole leaf, as in the *Viburnum lantana.*

VENTRICOSA *Spica* [*venter*, the belly] big-bellied; narrowing towards each extremity. *Ventricóſum Perianthium*, as in the *Eſculus. Ventricoſa Corolla*, as in the *Digitalis.*

VENTRICULOSUS *Calyx* [dim. à *venter*, the belly] bellying out in the middle, but not in ſo great degree as *Ventricoſus*; exemplified in the *Salicornia.*

VEPRECULÆ [dim. à *Vepres*, a brier or bramble] An order of plants in the *Fragmenta methodi naturalis* of Linnæus, in which are theſe genera, viz. *Rhamnus, Sideroxylum, Chryſophyllum, Lycium, Ceanothus, Philyca, Ceſtrum, Cateſbæa*, &c.

VERRUCOSA *Capfula* [*Verruca*, a wart] producing, on its furface, little knobs or warts, as in the *Euphorbia verrucofa*.

VERSATILIS *Anthera* [*verto*, to turn] when the *Anthera* is fixed horizontally, on the point of the *filamentum*, and confequently is fo poized, as to turn on it, like the needle of a compafs, as in the *Vitex*, *Linnæa*, *Geranium*, &c.

VERTICALIA *Folia* [*vertex*, the top of any thing] leaves fo fituated that their *bafs* is perpendicularly above their *apex* : applied only to aquatic plants.
Verticales flores, when the difk of the flower is turned as it were upfide down, facing the earth ; oppofed to *horizontales*.

VERTICILLATI *Rami*, *Flores*, *Folia* [à *Verticillum*, an axis or fpindle] branches, flowers, or leaves, furrounding the ftem like the *radii* of a wheel, *caulem annulatim ambientibus*. The fame as *Stellati*.
Verticillatæ, an order of plants in the *Fragmenta methodi naturalis* of Linnæus, containing thefe *genera*, viz. *Ajuga*, *Teucrium*, *Trichoftema*, *Thymus*, *Satureja*, &c. Verticillatæ funt fragrantes, nervinæ, refolventes, & pellentes : folia virtute pollent. *Lin.* The *Verticillatæ* are of

D d   the

the clafs and order, in the fexual fyftem, *Didynamia Gymnofpermia.*

Verticillata radix, a fpecies of the fibrous root exemplified in the aquatic and fenny plants. *Ludwig.*

VERTICILLUS [*vertex,* a whirlpool] a little whirl, axis, or fpindle ; a fpecies of inflorefcence in which the flowers grow in whirls, as in the *Marrubium.* A *Verticillus* may be either *feffilis, pedunculatus, nudus, involucratus, bractcatus, confertus,* or *diftans.*

VESICULA, a little bladder. The *Pericarpium* of the *Fucus.*

VESICULARIS *Scabrities* [*vefica,* a bladder] a fpecies of glandular *Scabrities,* roughnefs, fcarce vifible to the naked eye, refembling *veficulæ,* on the furface of fome plants, as in the *Mefcmbryanthemum, Aizoon, Tetragonia,* &c.

VEXILLUM, a ftandard ; the upright *petalum* of a papilionaceous *corolla.*

VILLOSUS *Caulis, Folium* [à *Villus,* wool] *pilis mollibus pubefcens ;* covered with diftinct but foft hairs ; woolly, as in the *Ulex europæus.*

VIRGATUS *Caulis* [*virga*, a rod] shooting forth straight slender branches, or rods, as in the *Artemisia campestris*.

VISCIDUM *Folium* [*viscus*, glue] when the surface of the leaf is clammy, *quod humore non fluido sed tenaci oblinitum*, as in the *Senecio viscosus*.

VISCOSITAS [*viscus*, glue] expresses that clamminess which covers the surface of some plants : it is ranged by Linnæus among the *Pubes*.

ULIGINOSA *Loca* [*Uligo*, the natural moisture of the earth] bogs ; *loca spongiosa, aqua putrida laborantia, colonis invisa, nec segetis, nec fœni proventui apta*.

UMBELLA [dim. ab *umbra*, a shadow] an umbel, or umbrella ; a *receptaculum* producing many equal *pedunculi* from one centre, as in the *Eryngium, Angelica, Cicuta, Pimpinella*, &c. An *Umbella* is either *simplex, composita, universalis*, or *partialis*.

UMBELLATUS *Flos*, properly so called, hath a common *receptaculum*, divided into *pedunculi* proceeding from the same point, a *germen* under the *corollula*, five distinct deciduous *Stamina*,

a

a bifid *piſtillum*, and two feeds united at their ſummits. They are of the claſs and order *Pentandria Digynia*.

Umbellatæ, a numerous order of plants in the *Fragmenta methodi naturalis* of Linnæus, a-mongſt which are the following *genera*, viz. *E-ryngium, Arctopus, Daucus, Angelica, Pimpinella*, &c. In ſiccis aromaticæ, calefacientes, & pellen-tes ; in aquoſis autem venenatæ ſunt : radice & ſeminibus pollent. *Lin.* The *Umbellatæ* conſtitute the ſeventh claſs in Tournefort.

UMBELLULA [dim. ab *Umbella*] the *Um-bella partialis* which diverges from the *apex* of each *pedunculus* of an *Umbella compoſita*.

UMBILICATUM *Folium* [ab *umbilicus*, a na-vel] See *Peltatum*.
Umbilicatus flos, faſhioned like a navel, as in the *Lichen miniatus*, &c.

UMBO ; Moriſon. See *Diſcus*.

UNANGULATUS *Caulis* [*Unus*, & *angulus*] forming one angle, as in the *Iris fœtidiſſima*.
UNCINATUM *Stigma* [*Uncinus*, an inſtru-ment hooked at the end] hooked, as in the *Viola lantana*.
Uncinata Ariſta, as in the *Geum urbanum*.

UN-

UNCTUOSUM *Folium,* clammy. Ludw. See *Viscidum,*

UNDATUM *Folium* [*unda,* a wave] or *Undulatum,* whose surface rises and falls in waves towards the margin, *cum discus folii versus marginem convexe adscendit & descendit,* as in the *Alchemilla, Potamogeton crispum.*

UNDULATA *Corolla* [*Undula,* dim. ab *unda,* a wave] waved, as in the *Gloriosa.*
Undulatum Folium, as in the *Oenothera mollissim.* Synon. with *Undatum.*

UNGUICULARIS *Caulis* [ab *Unguis,* a nail of the hand, &c] See *Unguis.*

UNGUIS [Ονυξ, *idem*] a nail of the hand or foot. The third degree in the Linnæan scale for measuring the parts of plants; the length of a finger-nail, containing six *Lineæ,* or half a Parisian inch. See *Mensura.*
Unguis, the *basis* of each *petalum* in a polypetalous *Corolla.*

UNICUS *Flos,* when the entire stem produces but one flower; different in signification from *Solitarius,* which see.

Unica

Unica Radix, a single root; having one bulb only, opposed to *duplicata*.

UNIFLORUS *Pedunculus* [*unus*, one, & *flos*, a flower] bearing one flower; having but one fructification on each *Pedunculus*.

UNILATERALIS *Racemus* [*unus*, one, & *latus*, a side] when the flowers grow only on one side of the *pedunculus*.

Unilaterales Cotyledones, growing on one side only; a species of the *Monocotyledones*, exemplified in *Palmæ*.

UNIVERSALIS *Umbella*, an universal umbel; the large *Umbella*, in an *Umbella composita*, to the extremities of whose *pedunculi* the *umbellulæ* are attached; opposed to *partialis*.

Universale Involucrum, when below the *Umbella universalis*.

VOLVA, the membranaceous *Calyx* of *fungi* : it may be *approximata*, or *remotissima*.

VOLUBILIS *Caulis* [à *volvo*, to roll] *spiraliter adscendens per ramum alienum* ; ascending spirally round the branch or stem of another. Their course is either *sinistrorsum*, to the left, with the sun's apparent motion, as in *Humulus*, *Helxine*,

Helxine, Lonicera, Tamus; or *dextrorfum*, the reverfe, as in *Convolvulus, Bafella, Phafeolus, Cy-
nanche, Euphorbia, Eupatorium.*

Volubilis Cirrhus, a twining tendril, *dextrorfum
.retrorfumque :* moft of the *leguminofæ* have ten-
drils of this kind.

URCEOLATA *Corolla* [*Urceolus*, ab *Urceus*, a
pitcher] bellying out like a pitcher, *pelvis inftar
inflata, et undique gibba.*

URENS *Caulis, Folium* [*Uro*, to burn] burn-
ing, ftinging, like nettles.

UTRICULI [ab *uter*, a bag, or bottle] *funt
vafcula repleta liquore fecreto ;* a fpecies of glandular fecretory veffels, obfervable in various
parts of the furface of fome plants, refembling
little bottles, replete with a fecerned liquor.

VULGARIS [*Vulgus*, the common people]
common. The trivial or fpecific name of many plants in the old botanifts, as the *Hydrocotyle
vulgaris*, &c. Synon. with *Frequens.*

F I N I S.